# INITIATION AND GROWTH OF EXPLOSION
# IN LIQUIDS AND SOLIDS

# INITIATION AND GROWTH OF EXPLOSION IN LIQUIDS AND SOLIDS

BY

F. P. BOWDEN & A. D. YOFFE

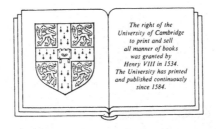

The right of the
University of Cambridge
to print and sell
all manner of books
was granted by
Henry VIII in 1534.
The University has printed
and published continuously
since 1584.

CAMBRIDGE UNIVERSITY PRESS

*Cambridge*

*London   New York   New Rochelle*

*Melbourne   Sydney*

CAMBRIDGE UNIVERSITY PRESS
Cambridge, New York, Melbourne, Madrid, Cape Town, Singapore, São Paulo, Delhi

Cambridge University Press
The Edinburgh Building, Cambridge CB2 8RU, UK

Published in the United States of America by Cambridge University Press, New York

www.cambridge.org
Information on this title: www.cambridge.org/9780521312332

First published in the Cambridge Monographs on Physics series 1952
Reissued in the Cambridge Science Classics series 1985
Re-issued in this digitally printed version 2009

A catalogue record for this publication is available from the British Library

Library of Congress Catalogue Card Number: 85-47912

ISBN 978-0-521-31233-2 paperback

# CONTENTS

## CHAPTER IV

## Initiation by Impact of Explosion in Solids

## CHAPTER V

## The Growth of Explosion to Detonation

# LIST OF PLATES

# PREFACE TO THE REISSUE

When Philip Bowden and I wrote this short monograph, published in 1952, we intended it as a self-contained book on a restricted topic and one which might stand the test of time. To a certain extent I believe this supposition was correct, and the main theme, that the initiation of explosion in condensed systems usually occurs in small local regions by a thermal (hot spot) mechanism, and that explosion can develop from this local region in a number of distinct stages, is generally accepted as being valid. The work described was carried out during the second world war and in the immediate post-war years when, among other factors, the hazards created by the presence of tiny gas bubbles or cavities in liquid or solid explosives were being examined. These turned out to have remarkably sensitizing effects, so that efforts were made to eliminate them for the safe handling of these unstable materials. However, this effect has now been put to good use, and modern industrial explosives of the emulsion or slurry type actually incorporate controlled size cavities in the form of hollow glass or plastic spheres of about 100 $\mu$m diameter.

Since this book was written, new techniques for studying rapid events have, as anticipated, been developed. In particular, improved high speed photographic equipment with excellent spatial resolution has been used with great success. In this connection, I would mention the experimental work in this laboratory directed by John Field and Munawar Chaudhri, since this is what I know best. They have examined in great detail the part played by the mechanical properties of solids in the initiation and growth of explosion which has helped in our understanding of the mechanisms involved. To assist the reader follow some of this later work, I have added a short list of references to reviews and books written since 1952, but this list is by no means exhaustive.

Philip Bowden died in 1968, but there can be little doubt he would have been delighted by the suggestion of a reissue

of this book in the Cambridge Science Classics series. My
own interests began to move into different areas of research
in the early sixties, but I do remember deriving considerable
enjoyment from writing this monograph. I hope it will
continue to be of interest and am very pleased it is being
reissued.

A. D. Yoffe

*Physics and Chemistry of Solids Laboratory*
*Cavendish Laboratory, 1985*

## References to reviews and books since 1952

*Fast Reactions in Solids*. F. P. Bowden and A. D. Yoffe, 1958, Butterworth Press, London.

Explosion in liquids and solids. F. P. Bowden and A. D. Yoffe, 1962, *Endeavour* **21**, 125.

The inorganic azides, A. D. Yoffe. In *Developments in Inorganic Nitrogen Chemistry*, Vol. 1, 1966, Ed. C. B. Colburn, Elsevier, Amsterdam.

*Detonics of High Explosives*. C. H. Johansson and P. A. Persson, 1970, Academic Press, New York.

*Initiation of Solid Explosives by Impact*. G. T. Afanas'ev and V. K. Bobolev, 1971, Israel Programme for Scientific Translations (Russian), Jerusalem.

*Energetic Materials*, Vols 1 and 2. Ed. H. D. Fair and R. F. Walker, 1977, Plenum Press, New York.

Proceedings of regular International Symposia on Detonation held in U.S.A. See, for example, 6th Symposium Annapolis, Maryland, 1981, published by U.S. Government Printing Office, Washington D.C.

Ignition mechanisms of explosives during mechanical deformation. J. E. Field, G. M. Swallowe and S. N. Heavens, 1982, *Proc. Roy. Soc.* A **382**, 231.

High speed photography of the interaction of shocks with voids in condensed media. M. M. Chaudhri, L. A. Almgren and A. Persson. In *Proceedings of the 15th International Congress on High Speed Photography and Photonics*, San Diego, 1982, ed. L. Engleman, S.P.I.E., Bellingham, WA.

# PREFACE

An account of an investigation into the physical processes involved in the sliding and impact of solids has been given in a recent monograph.* In addition to the physical changes, however, the act of rubbing or of striking the solids together may produce chemical changes and may initiate chemical reaction. The mechanism by which these changes are brought about is not always clear, and since the reaction is usually limited to the surface layer in the immediate region of contact, the amount of chemical decomposition is small and may escape detection. With explosives, however, this difficulty does not arise. The decomposition of a few neighbouring molecules on the surface can communicate itself to the bulk of the explosive so that detection of reaction is easy—sometimes only too easy. For this reason, during the past few years, we have been making a study of the decomposition by impact and by friction of explosive substances, and this monograph gives an account of these investigations. It deals mainly with the work of the writers and their collaborators; the approach is a direct experimental one and the work is still in progress, so that the picture presented is necessarily incomplete and elementary. Nevertheless, the results suggest that the mechanism of initiation is essentially simple. The latter part of the monograph describes a study of the way in which the tiny nucleus of decomposition can grow into a large-scale explosion.

The experimental work was carried out during the war years in the Tribophysics Division of the Australian Council for Scientific and Industrial Research Organization and since 1946 in Cambridge. We thank Sir David Rivett, F.R.S. and members of the Executive and Council of C.S.I.R.O. for interest and encouragement during the early stages of the work, and the Ministry of Supply (Air) for

* Bowden and Tabor (1950), *The Friction and Lubrication of Solids*. Oxford University Press.

grants made to the Cambridge Laboratory. Our indebtedness to others is great and the names of our research collaborators are given in italics in the text. Our thanks are due in particular to *Dr P. Gray* and *Dr O. A. Gurton* for their many helpful suggestions during the preparation of this monograph. Unless otherwise stated the experimental results quoted are due to the writers and their colleagues.

F. P. B.

A. D. Y.

*Research Laboratory on the Physics and Chemistry of Surfaces*
*Department of Physical Chemistry*
*Cambridge* 1951

## CHAPTER I

# INTRODUCTION

## 1.1. Scope of monograph

In the past many empirical tests have been devised to determine the sensitivity of explosive materials to shock, to impact and to friction. These practical tests are employed extensively and have been of value in deciding the various uses of explosives and the precautions necessary for their safe handling. It is true to say, however, that these practical tests do not always give a decisive answer. The behaviour of an explosive is, in a certain degree, unpredictable, in that explosions may occur under conditions which are normally considered to be 'safe'. Also the results have shed little light on the mechanism of initiation. It is an experimental inquiry into this question which forms the subject of this monograph. How is the mechanical energy of the blow or of rubbing able to produce the chemical excitation which leads to explosive reaction? One mechanism which has been proposed is that the reaction is 'tribochemical' in origin. That is to say that in some way the combined application of high pressures and the rapid shearing of adjacent molecular layers either causes a direct rupture of the molecule or alternatively produces sufficient deformation to bring about rapid chemical reaction. Since the bond energies involved may be of the order of 30–50 kcal., it is not easy to see how localized stresses of sufficient magnitude can arise.

The experiments described here suggest, however, that this is not the general mechanism. There is strong evidence that, for the majority of explosives, the initiation is thermal in origin. The mechanical energy must be degraded into heat and concentrated in a small region to form a hot spot. These hot spots, though small, are large compared with molecular dimensions, and they may perhaps be of the order of $10^{-5}$ to $10^{-3}$ cm. in diameter. Thermal ignition of the explosive then occurs at the hot spot.

It will be shown that the hot spots can be formed in three main ways:

(1) By the adiabatic compression of small entrapped bubbles of gas.

(2) By frictional hot spots formed: (a) on the confining surfaces, (b) on extraneous grit particles, (c) by intercrystalline friction of the explosive particles themselves.

(3) By viscous heating of the rapidly flowing explosive as it escapes from between the impacting surfaces.

Both theory and experiment suggest that the third method is operative only under extreme conditions, and that it is the first two which are of the greatest importance.

After a brief discussion of the formation of the explosion nucleus and the thermal theory of explosion (§§ 1.2, 1.3) we will consider first the initiation of explosion in liquids and solids by friction (Chapter II). Various physical methods for measuring the surface temperature of rubbing solids show that the local surface temperature may reach a high value even under comparatively gentle conditions of sliding. These temperature flashes which occur at the points of rubbing contact are of short duration, and the temperature rise is naturally limited by the melting-point of the rubbing solid. Experiment shows that the ignition of a liquid explosive such as nitroglycerine is due to the local hot spots formed on the surface of the sliding solids. Under the conditions of the experiment the temperature of the hot spots necessary to cause the explosion of nitroglycerine is about 480° C.

It is well known that the presence of grit particles renders many explosives more sensitive both to impact and to friction. This provides a method of introducing into the explosive hot spots of known maximum temperature, since the hot spot on the particle will, in general, be limited by its melting-point. Grit particles of known melting-point were therefore added to a variety of liquid and solid explosives, and their influence on friction sensitivity has been studied. It is found that the melting-point of the grit particle is of primary importance. With the solid secondary explosives (such as pentaerythritol tetranitrate or cyclonite) or with liquid explosives (such as nitroglycerine) all particles of melting-point greater than 430° C. are effective in causing explosion. Particles of melting-point below about 400° C. are ineffective. The primary explosives, such as lead azide, lead styphnate and mercury fulminate, also give similar results, although the limiting melting-point of the grit varies with the nature of the explosive and is somewhat higher. The effect of hardness, thermal conductivity and size of the grit particle and the difference between the intercrystalline friction of the primary and secondary explosives are discussed in § 2.4. The mechanism of initiation of explosion by impact is then discussed. Chapter III begins with liquid explosives, since these are homogeneous, and the variable factors, such as crystalline size, form and density of packing which influence the sensitivity of solid explosives, are absent. It is shown that the high impact sensitivity of liquid, gelatinous or plastic explosives is due to the presence, or to the trapping during impact, of minute gas bubbles. These tiny gas spaces or bubbles are heated by adiabatic compression and ignite the explosive. Under appropriate conditions these microscopic bubbles can confer an extreme sensitivity on the explosive so that it can be ignited by the gentlest blow. If precautions are taken to eliminate all bubbles the explosive is comparatively insensitive, and very high impact energies must be used. There is evidence that under these conditions the ignition is due to the viscous heating of the rapidly flowing explosive. Chapter IV describes experiments on the initiation of explosion in solids by impact. Again, it is shown that an important cause of initiation is the adiabatic compression of small gas spaces which are normally present between the crystals of the explosive or which are trapped during impact. Another source of hot spots is the presence of small grit particles, and again the melting-point of the particle is of primary importance. With the primary explosives hot spots can also be provided by friction of the explosive crystals. Some studies are also made of the time during the impact at which initiation is first observed.

Chapter V gives a brief account of the developments of some of the high-speed photographic methods and their application to the study of the way in which an explosion grows from the point of initiation and develops into a large-scale detonation. There is clear evidence that initiation begins at a hot spot, and a point of some interest is that frequently it does not grow at all; it is stillborn. Although there may be no outward sign of an explosion, sensitive photographic and other methods show that a micro-explosion has occurred and has died away without propagating to an appreciable distance. Moving-film studies show that in addition to being a precarious process the development of the explosion during the first few microseconds after its birth, or during the first few millimetres of its growth, is a complex one. With most explosives, both liquid and solid, the explosion begins as a comparatively slow and gentle deflagration which accelerates up to a speed of a few hundred metres per second and quickly passes over into a constant-velocity detonation which may be of the order of 2000 msec.$^{-1}$. It is suggested that the speed of burning in the first stage represents a mass movement of the gas products away from the point of initiation. The second stage of constant-velocity detonation may be identified with the *low*-velocity detonation observed in large charges. It is suggested that in solids and in viscous liquids the local hot spots can play just as important a part in the growth and propagation of the explosion as they do in initiation. Once again a common source of hot spots is the presence of small gas bubbles which are suddenly compressed by the shock front.

It is clear that the chemical reactions associated in the different physical stages of the explosion are themselves different, and analysis shows that the decomposition products formed during the early stages of the explosion correspond more closely to the thermal decomposition than they do to those formed by complete high-speed detonation.

There is evidence that the initiation of the explosion begins in the vapour phase, i.e. in the hot gas inside the bubble. For this reason, experiments on the thermal decomposition of the vapour of explosives such as nitroglycerine and methyl nitrate are described. The results show that reaction can occur when the concentration of the vapour is remarkably low. The normal ignition may be preceded by a 'blue glow' which is analogous to the cold flame in a hydrocarbon combustion. The results suggest that the propagation in the vapour phase is by a 'chain thermal' mechanism.

These general observations are of some practical interest, since they may help to explain the erratic but sometimes disastrous explosions which occur from time to time. They also suggest that the various practical 'tests' should be scrutinized closely and, if possible, the *cause* of initiation determined. If, for example, initiation is due to a hot spot on the confining surfaces the probability of an explosion can be diminished by making the surfaces of a low-melting alloy. If it is due to friction between particles it might be diminished by coating the particles with a lubricating wax. Neither of the methods would be effective, however, if the initiation were in reality due to an included bubble. There seems little doubt that, particularly in liquid explosives, the presence of even a single small bubble is a potential source of real danger. Complete removal is a matter of some difficulty, and, under certain conditions, the presence of a few tiny bubbles can cause initiation more readily than a considerable number of large ones over which the compression would be more uniformly divided. It is, perhaps, fortunate that the growth of the tiny explosion to one of appreciable dimensions is critical and requires favourable conditions.

## 1.2. The explosion nucleus

Several mechanisms exist by means of which a small nucleus of decomposition may develop into an explosion. The decomposition of all the explosive materials which will be considered is accompanied by the liberation of a great deal of energy as heat. If the rate of production of the heat is greater than the rate of heat loss by convection and conduction to the cooler surroundings, the reaction accelerates to a great speed and inflammation and explosion may occur. Such an explosion is called a 'thermal explosion'; the occurrence of such explosions is governed by the thermal properties of the system. A simple theoretical treatment of the factors involved was first given by van't Hoff (1884) and has been re-examined by Frank-Kamenetski (1939) and Rice (1940) and applied by Robertson (1947) to explosions in the condensed phase. The simple explanation of the thermal theory used by Semenoff (1935) and an outline of Robertson's account are given below in § 1.3.

Another way in which a reaction may be accelerated to an explosion occurs when one of the decomposition products acts as a catalyst. The catalyst or active intermediate is itself capable of reacting with undecomposed molecules to give the products of decomposition and be itself regenerated; if it does so with a lower energy of activation than the uncatalysed decomposition, the speed of the reaction may be greatly increased. For example, Robertson

(1948) finds that the ignition of liquid T.N.T. when heated to about 300° C. is facilitated by the presence of some partly decomposed T.N.T. The condition for explosion is then determined by the time required to build up a critical concentration of catalyst in the molten T.N.T.

Such catalysis by intermediates occurs in the radical chain mechanism which has been used to describe many explosive decompositions. Here the active intermediate is the free radical which may be formed in an initial endothermic fission reaction and which may then react, for example, with an undecomposed molecule to give further products. When more than one radical is produced from a reaction involving only one reactant-free radical, chain branching, is said to occur, and the reaction accelerates rapidly and explosion may result. Such explosions may occur under isothermal conditions. In many explosive reactions (which are known to accelerate by a branched radical chain mechanism) where the heat of reaction is large, it is impossible to maintain isothermal conditions, and self-heating does occur; such explosions are called 'chain thermal'. The explosions which are obtained with liquids such as nitroglycerine when heated to a high temperature are probably of this type.

## 1.3. Thermal theory of explosion

Before describing the main experimental investigations a simple account will be given in this section of the thermal theory of explosions.

In a chemical decomposition the rate of change for an $n$th order reaction occurring at $T°$ absolute is given by

$$\frac{dx}{dt} = k(a-x)^n, \tag{1.1}$$

where
$$k = Ae^{-E/RT}. \tag{1.2}$$

'$a$' is the initial concentration of reactant, $a-x$ the concentration at the time '$t$', '$A$' a constant and '$E$' the energy of activation. The thermal decomposition of nitroglycerine, for example, is first order, so that the velocity in this case is given by

$$\frac{dx}{dt} = A(a-x)e^{-E/RT}. \tag{1.3}$$

If $Q$ is the heat of reaction, the initial rate of liberation of heat $\dot{q}_1$ is

$$\dot{q}_1 = AaQe^{-E/RT}. \qquad (1.4)$$

This increases rapidly with temperature. Curves 1, 2 and 3 of fig. 1 show the dependence of $\dot{q}_1$ on $T$ for different values of $a$. The greater the initial concentration the higher the curve.

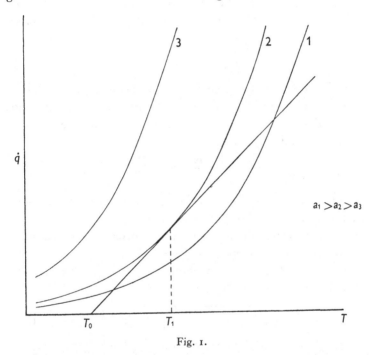

Fig. 1.

Heat will be lost both in warming the reacting mass and by conduction and convection to the surroundings. The rate of loss of heat $\dot{q}_2$ due to these effects is approximately proportional to the temperature excess $T - T_0$ of the explosive over its surroundings:

$$\dot{q}_2 = \underset{\substack{\text{conduction and}\\\text{convection}}}{K'(T - T_0)} + \underset{\text{self-heating}}{\Sigma C_v(T - T_0)} \qquad (1.5)$$

$$= K(T - T_0). \qquad (1.6)$$

The relation between $\dot{q}_2$ and $T$ is the straight line shown in fig. 1. In general this straight line cuts the heat production curve in two points or not at all. In the latter case the reaction once begun accelerates itself, since the heat loss is always less than gain. In the other case there is a range of temperatures for which a steady state is possible up to a certain limit beyond which acceleration again occurs. Critical conditions are represented by curve 2 where the curves for production and loss of heat touch. At this point

$$\dot{q}_1 = \dot{q}_2 \quad \text{and} \quad \ddot{q}_1 = \ddot{q}_2, \tag{1.7}$$

$$AaQe^{-E/RT} = K(T_1 - T_0), \tag{1.8}$$

and
$$\frac{AaQE}{RT_1^2} e^{-E/RT_1} = K. \tag{1.9}$$

Semenoff found that the value for $K$ calculated by equation (1.9) for nitroglycerine corresponds to the thermal conductivity, and concludes that the explosions produced during heating are thermal.

From (1.8) and (1.9) we get

$$T_1 - T_0 = \frac{RT_1^2}{E} \simeq \frac{RT_0^2}{E}. \tag{1.10}$$

Equation (1.10) represents the self-heating necessary for explosion and amounts to $c.$ $10°\,\dot{C}.$ for nitroglycerine.

For solid explosives Robertson (1947) has extended the treatment given by Frank-Kamenetski for gases. For an infinite slab of explosive of width $2d$ at $T°$ above the vessel temperature $T_0$, the general expression for heat flow is

$$c\rho\,\frac{\partial T}{\partial t} = K\left[\frac{\partial^2 T}{\partial x^2} + \frac{\partial^2 T}{\partial y^2} + \frac{\partial^2 T}{\partial z^2}\right] + Q'Ae^{-E/R(T_0+T)}, \tag{1.11}$$

self-heating      heat-loss by conduction      rate of production of heat by chemical reaction

where $c$ is the specific heat, $\rho$ the density, and $Q'$ the heat of reaction per unit volume.

For a steady state $(\partial T/\partial t = 0)$ to be reached and considering heat flow in the $X$-axis, the equation reduces to

$$K\frac{\partial^2 T}{\partial x^2} + Q'Ae^{-E/R(T_0+T)} = 0. \tag{1.12}$$

The maximum value $T_m$ for the temperature of the explosive above the surroundings is given by

$$d\,\sqrt{\left(\frac{Q'}{K}\,Ae^{-E/RT_0}\,\frac{E}{2RT_0^2}\right)}\,e^{(E/2RT_0^2)\,T_m} = 1\cdot2, \tag{1.13}$$

where $2d$ is the width of the explosive slab. The critical value for a non-dimensional parameter $P$ when explosion occurs is

$$P = d^2 \frac{Q'}{K} A e^{-E/RT_0} \frac{E}{RT_0^2} = 0.88. \qquad (1.14)$$

When the expression $P$ is greater than $0.88$ explosion should occur. For P.E.T.N. and cyclonite, Robertson obtained values for $P$ between 5 and 10. This discrepancy is explained in terms of loss of heat by convection which are not considered in the above treatment as well as by conduction.

A simple derivation is given in Chapter IV for the growth of explosion from a small hot spot in the explosive. Such hot spots may be produced during impact and friction. Under these conditions, however, the time involved is very short and heat will be lost mainly by conduction, a condition favourable for the application of the Frank-Kamenetski treatment.

## 1.4. Size of nucleus of decomposition

Garner (1938) and Muraour (1938) have developed some ideas concerning the size which the small nucleus of decomposition must have before it will develop into explosion. The basis of this theory is that self-heating occurs in a localized spot in the explosive and that the explosion grows from this region. When lead azide is heated at 290° C. in a vacuum ignition occurs after an induction period of about 20 sec. Garner considers that explosion will follow when two adjacent groups decompose simultaneously (within $10^{-13}$ sec.). The calculations are based on the following considerations.

In the case of lead azide, the activation energy $E$ for the thermal decomposition is 38 kcal., and the heat of reaction $Q$ is 106 kcal./mole. The energy set free during the decomposition of one molecule is $Q + E$ or 144 kcal. This is sufficient to activate the decomposition of $\left( \frac{Q+E}{E} \simeq 4 \right)$ three to four neighbouring $PbN_6$ molecules (i.e. six to eight $N_3$ ions). However, the number of nearest neighbours about each $PbN_6$ molecule is greater than this, which means that the 144 kcal. liberated are insufficient to set up a spherical wave of reaction. However, if two adjacent $PbN_6$ molecules decompose simultaneously, sufficient energy is liberated to activate fifteen surrounding ions. This, according to Garner, is sufficient to start a spherical wave of reaction which ultimately develops into a detonation wave.

Two molecules are considered as decomposing simultaneously if they do so within $10^{-13}$ sec., which is the lowest frequency of vibration of the $N_3$ ion. Garner has made some calculations from his thermal decomposition data on the probability of the simultaneous decomposition of two and three adjacent molecules, and concludes that the binary event is the cause of the initiation of explosion.

Kallman and Schrankler (1933) and Muraour (1933) have obtained results in support of this view. They find that when rapidly moving atomic and nuclear projectiles are used, that $\alpha$-particles and electrons which possess small diameters do not give rise to detonation, but cause molecules to decompose singly. When ions of argon and mercury are used which are large enough to activate a group of neighbouring molecules, explosion does occur.

Hawkes and Winkler (1947), on the other hand, consider that there is no need to postulate the simultaneous decomposition of two molecules, but that all the phenomena observed during the explosion of lead azide may be explained simply by a self-heating mechanism. Other criticisms have been put forward against Garner's theory by Andreev and Chariton (1934) and Macdonald (1938).

Andreev (1934) has pointed out that three factors must be considered in the growth of the explosion. During the decomposition of a molecule of explosive, the energy of each of the product molecules is, on the simplest view, given by

$$q = \frac{Q+E}{n}, \qquad (1.15)$$

where $n$ is the number of product molecules formed. The tendency for the decomposing molecule to activate another will increase with increase in the ratio $q/E$. This mechanism visualizes the activation of explosive molecules by 'hot' product molecules. The factors favouring the propagation of the reaction will thus be a high heat of reaction, a low activation energy for decomposition and a small number of product molecules.

## 1.5. Direct mechanical breaking of molecules

When explosives are detonated by a mechanical blow, the results can best be explained by a thermal mechanism. However, some decompositions have been observed among non-explosive materials where it seems the molecules are broken directly without the conversion of the mechanical energy to heat. Such decompositions may be termed 'tribochemical'. These are usually connected with the breakdown of high polymers into low molecular weight fractions under the action of shearing forces. Hargrave (1947) observed the irreversible breakdown of polyisobutylene when a solution in hexane was stirred. Similar effects have been reported by Schmid (1940), who observed depolymerization in solution due to the action of ultrasonic vibrations. The action of ultrasonic vibrations on some sensitive explosive materials has also been tried. Bobolev and Chariton (1937) found that nitrogen chloride was decomposed by this means. Nitrogen iodide can also be exploded by ultrasonic vibrations when the nitrogen iodide is immersed in a liquid. However, Marinesco (1935) has suggested that the explosion of the nitrogen iodide is due to the adiabatic heating of small air bubbles adhering to the crystals by the pressure pulses created during the passage of the ultrasonic vibrations.

There are several other observations where the thermal theory cannot be applied directly. It has been shown, for example, that crystals of lead azide when growing in solution will detonate spontaneously. Other effects, such as the exposure of a fresh surface, may be important. For example, Meldrum (1940) has shown that nitrogen iodide explodes at room temperature if ammonia is removed from the surface. Eggert (1921) has also shown that the application of a gas pressure of 300 atm. will detonate nitrogen iodide. It is difficult to see how hot spots could be responsible for initiations of this type. Another important factor, which, under certain conditions, may be responsible for the initiation, is the development of electric charges on breaking or moving crystals. It is well known that the discharge of very small amounts of static electricity may detonate an explosive. For a general review on tribochemistry see *Bowden* and *Yoffe* (1948).

## CHAPTER II

# INITIATION OF EXPLOSION BY FRICTION

When solid surfaces are placed together, contact will occur only locally at the summits of the highest irregularities so that the real area of contact is, in general, very small. If now the solids are rubbed together the frictional energy is dissipated mainly in the form of heat and is concentrated in these regions so that the temperature rise at the surface of the local points of contact may be quite high. A general account of the measurement of surface temperature has been given recently by *Bowden* and *Tabor* (*The Friction and Lubrication of Solids*, Oxford University Press). If the hot spots are formed by friction on a solid surface there is evidence that the maximum temperature rise is usually limited by the melting-point of the surface. By using surfaces of known melting-point it is therefore possible to fix the transient temperature that can be generated by friction. *Bowden* and co-workers used this method to determine the conditions for the initiation of explosion in liquids and solids by friction. Before describing these experiments, it will be useful to discuss the formation of hot spots on rubbing surfaces, and then to show the correlation that exists between the formation of hot spots and the incidence of explosion.

## 2.1. Surface temperature of sliding metals

When two dissimilar metals are rubbed together the surface temperature may be measured by using the sliding contact as a thermocouple (*Bowden* and *Ridler*, 1936; *Bowden, Stone* and *Tudor*, 1947). A measurement of the electromotive force generated on sliding then provides a record of the surface temperature. The temperatures reached depend on the load, speed and thermal conductivity of the metals, but even under comparatively gentle conditions of sliding the temperature may be high. The maximum temperature reached is limited by the melting-point of the metal. With high-melting metals temperatures of the order of $1000°$ C.

are attained. These temperatures are confined to the surface layer. Also the temperature fluctuates very rapidly during sliding, and it is necessary to use an instrument of rapid response such as a cathode-ray oscillograph in order to record them (see fig, 2). It is clear that the very high temperature flashes of 1000° C. may last for only a few ten-thousandths of a second. Even if the surfaces are flooded with water or other liquids, the temperature during rubbing is still high.

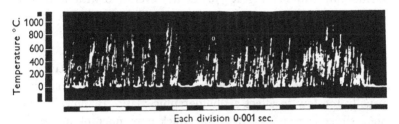

Each division 0·001 sec.

Fig. 2. Cathode-ray trace of thermal e.m.f. developed between a constantan slider and a lapped steel surface. Load 500 g. Speed of sliding 300 cm.sec.$^{-1}$. The temperature flashes are high and of very short duration.

## 2.2. Surface temperature of rubbing solids; non-metals

With non-metallic solids such as glass or quartz, the thermal conductivity is low compared with that of metals, and we should expect therefore that the local hot spots would occur more readily. The temperature rise $T - T_0$ is given approximately by the expression

$$T - T_0 = \frac{\mu W V}{4aJ} \frac{1}{k_1 + k_2},$$ (2.1)

where $\mu$ is the coefficient of friction, $W$ the load between the surfaces, $V$ the velocity of sliding, $a$ the radius of the circular region of contact, $J$ the mechanical equivalent of heat, $k_1$ and $k_2$ the thermal conductivities of the rubbing solids. Unfortunately, the thermocouple method cannot be used with these solids, but it is possible to see the hot spots. If polished surfaces of glass or quartz are used and the apparatus so arranged that a clear image of the rubbing surfaces can be seen, it is found that when sliding starts, a number of tiny stars of light appear at the interface between the rubbing surfaces. The points of light are reddish in colour at low speeds and become whiter and brighter as the speed or load is increased. These correspond to small hot spots (diameter $10^{-2}$ to $10^{-4}$ cm.) on the surface, and their position and distribution over the surface change from instant to instant. By making one of the surfaces of metal it is possible to fix approximately the temperature at which hot spots first become visible. If metals or alloys melting below 520° C. are slid on glass or quartz no hot spots can be seen even at the highest speeds and loads. With a gold aluminium alloy melting at 570° C. however, and with all metals above this, the hot spots are readily seen. This fixes the temperature at which the hot spots first become visible to the eye at between 520 and 570° C.

(1) *Photographic recording of hot spots.* Although the occurrence of these transient hot spots is readily observable, the intensity is usually too low to affect a photographic plate. If, however, the steel slider is run over the same track a number of times, the cumulative light from the hot spots is enough to produce a record on a photographic plate held under the rotating glass disk. The surfaces are allowed to run for 2 min. in a circular track, and the result obtained is shown in fig. 3. The circles of decreasing radius mean decreasing sliding speed. The lowest speed at which the hot spots are recorded under the conditions of the experiment is 70 cm.sec.$^{-1}$. Visible hot spots ($T \simeq 570°$ C.) may be seen when the speed of sliding is as low as 30 cm.sec.$^{-1}$.

(2) *Properties of transient hot spots.* Another method for studying the occurrence of transient hot spots on transparent materials such as glass and quartz is by the use of infra-red sensitive photocells such as the lead sulphide cell (*Thomas*, 1949). These cells can be constructed with very small time constants, so that transient hot spots of very short duration can be detected and measured. In addition, the cells are capable of responding to radiations of wavelength greater than the visible spectrum, so that hot spots of comparatively low temperatures can be detected. For metals sliding on glass it is shown that the area of some of the hot spots may be $c$. $10^{-3}$ cm.$^2$, and their duration $c$. $10^{-5}$ to $10^{-3}$ sec.

(3) *Effect of thermal conductivity on incidence of hot spots.* When metals of different thermal conductivity are allowed to slide on a glass disk, the results obtained for the incidence of visible hot spots

are shown in fig. 4. The sliding speed is kept constant and the load varied until hot spots just become visible. At the same time the coefficient of friction is measured so that the frictional force necessary to give hot spots at any given speed can be determined. It will be seen from fig. 4 that hot spots occur more readily the lower the thermal conductivity of the slider. For example, at a surface speed of 110 cm.sec.$^{-1}$ a tungsten slider gives hot spots when the frictional force is 2600 g., whilst with a constantan slider

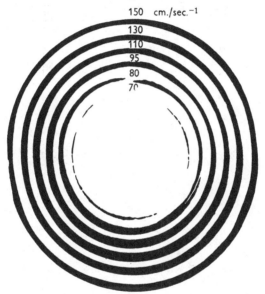

Fig. 3. Photographic record of hot spots produced by a steel surface sliding on a rotating glass plate. Load 1200 g. Coefficient of friction 0·6. The sliding speeds are given. The innermost track (speed 70 cm.sec.$^{-1}$) represents the lowest speed at which hot spots are recorded on the photographic plate.

hot spots occur when the frictional force is only 350 g. If the surfaces are flooded with a liquid such as a glycerine-water mixture, the hot spots still occur and the results for metals of varying thermal conductivity are similar to those given in fig. 4. The main difference is that all the curves are shifted upwards and a higher frictional force (about seven times as great) is necessary to produce hot spots when the surfaces are flooded. This difference is considerable, but it is clear that the presence of the liquid film is not able to prevent the occurrence of high local temperatures.

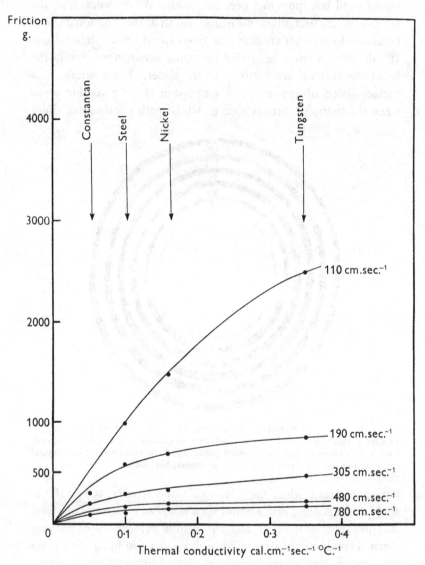

Fig. 4. The generation of visual hot spots on a clean glass surface. The vertical ordinate gives the frictional force at which hot spots appear for various sliding speeds, using sliders of constantan, steel, nickel, and tungsten.

## 2.3. Initiation by friction of explosion in liquids

(1) *Effect of thermal conductivity.* The same apparatus as that described in § 2.2 may be used, namely, a lower rotating disk of glass, and metal sliders of different thermal conductivity. The glass

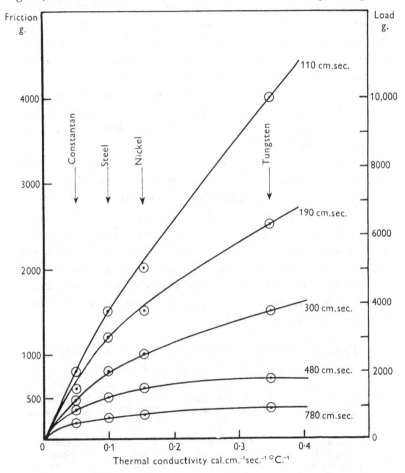

Fig. 5. The explosion of nitroglycerine by friction for sliders of constantan ($k=0\cdot05$), steel ($k=0\cdot1$), nickel ($k=0\cdot16$), and tungsten ($k=0\cdot35$) rubbing on a glass surface. This figure should be compared with fig. 4.

disk is now covered with a thin film of a liquid explosive such as nitroglycerine and is revolved at a fixed speed, and the load on the slider is increased until explosion occurs. In some experiments explosion occurs immediately, in others there is a time interval which varies from a few seconds to half a minute. In spite of this variation, the frictional force and speed at which explosion occurs are fairly well defined and reasonably reproducible. Fig. 5 gives the results obtained. Each of the curves is for a constant speed of revolution and shows a plot of the minimum frictional force to produce explosion against the thermal conductivity of the slider. These curves should be compared with those given in fig. 4, which are plots of the slider conductivities against friction necessary to give visible hot spots on the surface.

The curves in fig. 5 are very similar in form to those given in fig. 4. It is clear that with a poor thermal conductor such as constantan, explosion occurs more readily than with a good conductor such as tungsten.

(2) *Temperature of hot spot necessary for explosion.* If the melting-point of the slider is altered, it is again possible to determine the temperature necessary to produce explosion. Sliders are made from a range of metals and alloys selected so that they do not oxidize readily even at elevated temperatures. The lower surface is of glass and is covered with a layer of nitroglycerine, a slider of the appropriate metal is used and the load or the speed is increased until explosion occurs. It is found that with metals or metallic alloys melting below 450° C. no explosion occurs even at very high load and speeds (see Table I).

TABLE I. *Incidence of hot spots and explosions with metal sliders of different melting-points on glass*

| Alloy composition | Melting-point (° C.) | Hardness (Vickers) | Visual hot spots on clean glass | Explosions of nitroglycerine |
|---|---|---|---|---|
| 80 Au, 20 Sn | 300 | 230 | None | None |
| 80 Au, 20 Pb | 420 | 108 | None | None |
| 75 Au, 25 Te | 450 | 120 | None | None |
| 73 Ag, 27 Sn | 480 | 93 | None | Several |
| 70 Ag, 30 Sb | 480 | 120 | None | Several |
| 80 Ag, 20 As | 500 | 170 | None | Several |
| 50 Au, 50 Cd | 520 | — | None | Several |
| 92 Au, 8 Al | 570 | 221 | Readily | Readily |
| Constantan | 1200 | 130 | Readily | Readily |
| Ni | 1450 | 170 | Readily | Readily |
| Fe | 1500 | 130 | Readily | Readily |
| W | 3000 | — | Readily | Readily |

An alloy melting at 480° C. and all metals melting above it, however, cause explosions. This indicates that the local high-surface temperature which is just able to produce explosions under these conditions is about 480° C. This surface temperature is slightly lower than that necessary to give visible hot spots (520–570° C.) (see Table I).

(3) *Thermoelectric measurement of surface temperature at explosion.* When the lower rotating surface is a steel disk and the slider of constantan, the temperature at the junction may be estimated by the thermoelectric method described earlier. If a thin film of nitroglycerine is placed on the steel disk explosion occurs at a given load and speed, and the light from the explosion may be picked up by a photoelectric cell (see fig. 6). The incidence of explosion may be recorded on the same trace as the temperature flashes. Fig. 7 shows such a trace of the local temperature flashes and the incidence of explosion at *A*. Again the temperature of the hot spots required to initiate explosion is in the region of 500° C.

**2.4. Initiation of explosion in solids. The influence of grit**

It is well known that the presence of small quantities of grit increases the friction and impact sensitivity of solid explosives. Quantitative experiments on this were carried out by Taylor and Weale (1938), who studied the effect of glass and of carborundum particles on the impact sensitivity of a number of explosives. At the time they suggested that the initiation of the explosion was due to some tribochemical mechanism, and it is usually considered that it is the *hardness* of the grit particle which is the most important factor. From the experiments described above, however, it would seem more probable that the sensitization by grit and the initiation of the explosion is due to the formation of hot spots on the grit particles. If this is so we should expect *the melting-point* of the grit particle to be of primary importance. In fact, the addition of grit particles of known melting-point would be a convenient method of introducing into the explosive hot spots of known maximum temperature, since, as we have seen, the temperature rise is limited by the melting-point of the solid. A series of experiments was carried out (*Bowden* and *Gurton*, 1948*b*) to investigate this point.

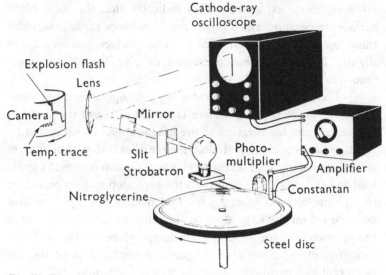

Fig. 6. Diagrammatic representation of the apparatus used to record surface temperature and the incidence of explosion.

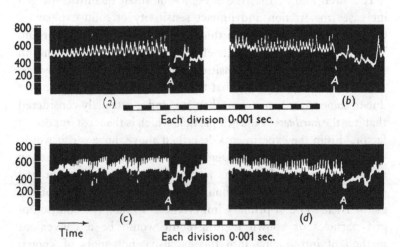

Fig. 7. Cathode-ray trace of thermal e.m.f. during sliding of a constantan pin on a steel surface covered with nitroglycerine, using apparatus shown in fig. 6. Explosion occurs at vertical lines *A*.

The apparatus used is shown in fig. 8 and is designed so that a thin layer of solid explosive is subjected to rapid shear while held under a known load. The sliding bar is set in motion by allowing the pendulum hammer to fall from a suitable height. About 25 mg. of explosive are used in these experiments. (1) *Results with secondary explosives.* When P.E.T.N. (penta erythritol tetranitrate) is subjected to this experiment no explosions are obtained when the highest loads and rates of shear are employed. However, if a few small particles of glass are introduced into the explosive before the test is carried out, explosion is obtained in every experiment. The initial size of the particles used is of the order of 100 $\mu$. This experiment was repeated with other 'grits' of known melting-points, the materials used as grit being crushed minerals and inorganic salts. The minerals chosen were among the softer ones (hardness on the Mohs' scale 2–3·5), so that the difference in hardness between most of the materials used was not very great.

Table II gives the explosion efficiencies* obtained with a number of grits together with the hardness and melting-points of the materials. For convenience the impact results, which will be discussed in Chapter IV, are included. If the friction explosion efficiencies set out in column 4 are compared with the hardness values in column 2, no correlation can be observed, but if they are compared with the melting-point of the grits as set out in column 3, a remarkably sharp division is apparent. All grits of melting-point greater than 430° C. are effective in causing explosion, while all grits less than about 400° C. are ineffective. Since the highest temperature reached by rubbing two solids together is limited by the melting-point of the lower melting solid, it follows that the hottest spot on a piece of grit rubbed on steel, or on another piece of grit, will not, in general, have a temperature above the melting-point of the grit. Thus these experiments show that when hot spots of 430° C. and upwards are produced in P.E.T.N. explosion usually follows. In the absence of any hot spot greater than 400° C. no explosion occurs at the highest rates of shear used in these experiments.

Cyclonite gives results which are essentially similar to those obtained with P.E.T.N. Again grits of high melting-point are

* For definition of explosion efficiency see p. 29.

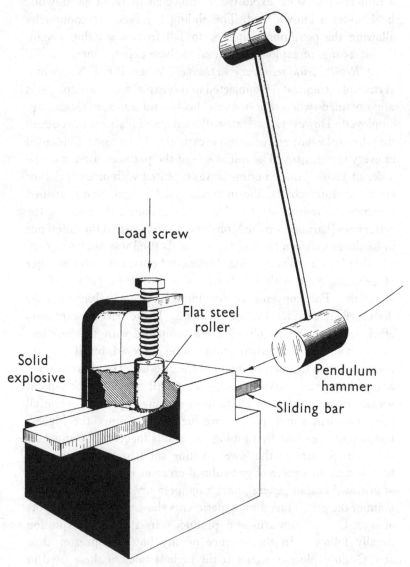

Fig. 8. Illustration of apparatus used for determining the friction sensitivity of solid explosives.

much more effective than grits of low melting-point, and no
explosions are obtained if the explosive is pure, or mixed with any
grit of melting-point less than 400° C.

TABLE II. *Initiation of explosion by friction of P.E.T.N.*
*in the presence of grit*

| Grit added | Hardness (Mohs' scale) | Melting-point (° C.) | Friction explosion efficiency (%) | Impact explosion efficiency (%) |
|---|---|---|---|---|
| Nil (pure P.E.T.N.) | 1·8 | 141 | 0 | 2 |
| Ammonium nitrate | 2–3 | 169 | 0 | 3 |
| Potassium bisulphate | 3 | 210 | 0 | 3 |
| Silver nitrate | 2–3 | 212 | 0 | 2 |
| Sodium dichromate | 2–3 | 320 | 0 | 0 |
| Sodium acetate | 1–5 | 324 | 0 | 0 |
| Potassium nitrate | 2–3 | 334 | 0 | 0 |
| Potassium dichromate | 2–3 | 398 | 0 | 0 |
| Silver bromide | 2–3 | 434 | 50 | 6 |
| Lead chloride | 2–3 | 501 | 60 | 27 |
| Silver iodide | 2–3 | 550 | 100 | — |
| Borax | 3–4 | 560 | 100 | 30 |
| Bismuthinite | 2–2·5 | 685 | 100 | 42 |
| Glass | 7 | 800 | 100 | 100 |
| Rock salt | 2–2·5 | 804 | 50 | 6 |
| Chalcocite | 3–3·5 | 1100 | 100 | 50 |
| Galena | 2·5–2·7 | 1114 | 100 | 60 |
| Calcite | 3 | 1339 | 100 | 43 |

(2) *Results with primary explosives.* Experiments have been
carried out on this apparatus with four initiating explosives,
namely, lead azide, lead styphnate, tetrazene and mercury fulminate, and there is one marked difference in the results. When these
explosives are subjected to the stringent conditions of test used in
the experiments with P.E.T.N. and cyclonite, they explode in the
absence of any grit. In consequence, the load applied was reduced
to 64 kg., and the maximum height of fall of the pendulum to give
no explosions was determined. Table III shows the relative
sensitivities of the four explosives.

A fundamental difference between these explosives and the
secondary explosives (P.E.T.N. and cyclonite) is shown in their
behaviour when heated slowly. The secondary explosives melt
before decomposing, while the primary explosives investigated
cannot be melted. They decompose explosively *while they are still*

*in the solid* state. It is likely, therefore, that hot spots which develop on the surfaces of these explosive crystals, or between the crystals and the steel surfaces, could reach temperatures above the minimum necessary for explosion. With the secondary explosives this will not occur, since the temperature rise is limited by the melting of the explosive.

TABLE III. *Initiation of explosives by friction in the absence of grit*

| Explosive | Melting-point (°C.) | Load (kg.) | Height of fall (cm.) | Explosion efficiency (%) |
|---|---|---|---|---|
| P.E.T.N. | 141 | 1600 | 70 | 0 |
| Cyclonite | 200 | 1600 | 70 | 0 |
| Lead azide | >335 | 1600 | 70 | 100 |
|  |  | 64 | 70 | 10 |
|  |  | 64 | 60 | 0 |
| Lead styphnate | >250 | 64 | 60 | 80 |
| , |  |  | 45 | 60 |
|  |  |  | 40 | 0 |
| Mercury fulminate | >145 | 64 | 5 | 10 |
|  |  |  | 2½ | 0 |

The effect of added grits on two of these explosives has been studied, and again it is found that only grits of melting-point 500° C. and greater are effective sensitizers for lead azide and lead styphnate. Some results for lead azide and lead styphnate are set out in Table IV.

TABLE IV. *Initiation of explosion in lead azide and lead styphnate in the presence of grit*

Load 64 kg.

| Grit | Hardness (Mohs' scale) | Melting-point (°C.) | Explosion efficiency (%) | |
|---|---|---|---|---|
|  |  |  | Lead azide (height of fall 60 cm.) | Lead styphnate (height of fall 40 cm.) |
| Nil | — | — | 0 | 0 |
| Silver nitrate | 2–3 | 212 | 0 | 0 |
| Silver bromide | 2–3 | 434 | 0 | 3 |
| Lead chloride | 2–3 | 501 | 30 | 21 |
| Silver iodide | 2–3 | 550 | 100 | 83 |
| Borax | 3–4 | 560 | 100 | 72 |
| Bismuthinite | 2–2·5 | 685 | 100 | 100 |
| Chalcocite | 3–3·5 | 1100 | 100 | 100 |
| Galena | 2·5–2·7 | 1114 | 100 | 100 |
| Calcite | 3 | 1339 | 100 | 93 |

Mercury fulminate and tetrazene were not treated in this way, since they were considered to be too sensitive to friction even in the absence of grit. However, when mercury fulminate and tetrazene are mixed with a small amount of grit (*c.* 7%) and subjected to impact between two flat surfaces, again a clear-cut relation is obtained between the melting-point of the particles and its influence on sensitivity. Grits melting above 500° C. have a marked effect in increasing the sensitivity (see Chapter IV).

(3) *Influence of size and thermal conductivity of the grit particles.* We should expect that the size, the hardness, and thermal conductivity of the grit particle will all play an important part because of the influence they have on the ease of the formation of hot spots. If the particles are too small and too numerous the energy of impact is dissipated over many points of contact distributed over a large area, so that no single spot reaches the required temperature for ignition. Experiments on the frictional initiation of nitroglycerine rubbed between metal surfaces in the presence of carborundum have, in fact, shown that larger grit particles (100$\mu$ in size) are more effective in producing hot spots than smaller particles (0·6–10$\mu$).

The general relation between thermal conductivity, the formation of hot spots, and the incidence of explosion has already been established with nitroglycerine. An indication of the importance of the thermal conductivity of the grit particle is provided by rock salt. The hardness of rock salt is 2–3·5 Mohs and its melting-point is *c.* 800° C. It should therefore sensitize the explosives to impact and friction. Experiments show that this is the case with P.E.T.N. and lead styphnate. In general, however, the explosion efficiency is appreciably lower than with the other grit particles of similar melting-point but lower thermal conductivity.

(4) *Influence of the hardness of the grit.* It is clear that although the maximum hot-spot temperature is fixed by the melting-point of the particle, the ease with which the hot spot is formed is very dependent upon the hardness. With a hard sharp particle the stresses are concentrated at one or two points so that it requires a much smaller energy, under conditions both of impact and friction, to produce a localized temperature rise of the necessary magnitude. If the particle is soft it is plastically deformed or

crushed so that this local concentration of the energy is not possible. For this reason we should expect that hard particles would be more effective than soft ones, *provided the melting-point of the particles is above the critical value.* Experiments carried out with a wide range of explosives and of grit particles show that this is true. Copp and Ubbelohde (1948) have also studied the friction sensitivity of explosives in the presence of grit. They did not investigate the influence of the melting-point, but they have stressed the importance of the hardness of the grit.

(5) *Transient temperature for explosion.* All these experiments show that, for a number of liquid and solid explosives, the frictional hot spot must have a minimum temperature which is estimated to be between 400 and 500° C. This temperature is well above the ignition temperature determined when explosives are heated in a furnace. In the case of nitroglycerine this latter temperature is about 200° C. when the nitroglycerine is heated in bulk in an open vessel. Roginsky (1932) was able to show that below 200° C. there was a time lag before explosion given by $\tau = 1\cdot 5 \times 10^{10}\, e^{25000/RT}$. The ignition temperature is not a characteristic property of an explosive but varies with the mass of explosive, the rate of heating and dimensions of the apparatus.

The hot spots generated by friction are transient and last for a very short time, which depends on the experimental conditions, but which is usually of the order of $10^{-3}$ to $10^{-5}$ sec. For this reason the hot-spot temperature for explosion is higher than conventional ignition temperatures. An extrapolation of Roginsky's data to 500° C. gives a value of $c$. $10^{-4}$ sec. for the time lag before explosion, and this agrees well with the duration of hot spots determined experimentally.

Copp and Ubbelohde (1948) have estimated hot-spot temperatures in a different way. With most explosives the thermal decomposition obeys a first-order law, such that the fraction decomposed $\alpha$ is related to the velocity constant $k$ at $T°$ K. by $k = \dfrac{1}{t}\ln\dfrac{1}{1-\alpha}$. In order that the explosion may propagate, a finite quantity of the explosive must decompose within the time of application of the hot spot. Since $k$ for many explosive decompositions is known, e.g.

Nitroglycerine $\quad k = 10^{20 \cdot 5} \quad e^{-48000/RT}$ (Roginsky, 1932)

P.E.T.N. $\qquad k = 10^{19 \cdot 8} \quad e^{-47000/RT}$ (Robertson, 1948 $b$)

Mercury fulminate $\quad k = 10^{11 \cdot 05} \quad e^{-25400/RT}$ (Vaughan and

Phillips, 1949)

it is possible to obtain values for $\alpha$ as a function of the explosive temperature $T^\circ$ K. and time $t$. A few results for P.E.T.N. are given in Table V.

TABLE V. *Fraction $\alpha$ of P.E.T.N. decomposed for different time intervals t*

| Temperature | | Time interval $t$ | | |
|---|---|---|---|---|
| ° K. | ° C. | $10^{-3}$ sec. | $5 \times 10^{-4}$ sec. | $10^{-5}$ sec. |
| 500 | 227 | 0·0003 | 0·0001 | 0·0000 |
| 600 | 327 | 0·47 | 0·27 | 0·006 |
| 700 | 427 | 1·0 | 1·0 | 0·82 |
| 750 | 477 | 1·0 | 1·0 | 1·0 |

Since the time of duration of the hot spots is between $10^{-3}$ and $10^{-5}$ sec., hot-spot temperatures of 400–500° C. are required, which is in reasonable agreement with the values determined experimentally by *Bowden* and *Gurton*. Since this calculation involves extrapolations outside the range of temperature over which determinations of the velocity constant have been carried out, it must be applied with reserve. It will be seen, however, that it does give a result of the right order.

CHAPTER III

# INITIATION BY IMPACT OF EXPLOSION IN LIQUIDS

The usual method of initiating, by impact, an explosion in a liquid or solid is to place some explosive on a flat anvil and hit it with a flat hammer. In order to make the experiment more reproducible an apparatus similar to that shown in fig. 9 may be used, or, alternatively, the explosive may be confined between two

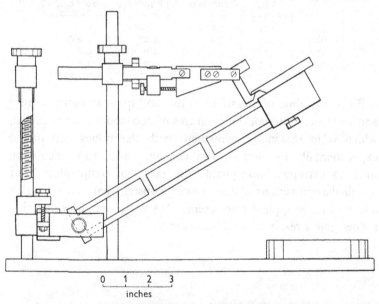

0  1  2  3
inches

Fig. 9. Pendulum hammer apparatus.

steel rollers and a freely falling ball allowed to drop on to the top roller. Explosion occurs if the energy of the impact is sufficiently high, and the sensitivity of the explosive may be expressed in terms of the potential energy of the striker necessary to produce explosion (i.e. product of the mass of the striker in g. multiplied by the height of fall in cm.). Most observers find that the incidence of explosion is to some degree erratic. Initiation may or may not occur, and it is usual to carry out a statistical analysis involving a large number of experiments at different heights. The energy required to give an 'explosion efficiency'

$$\left( \text{i.e.} \ \frac{\text{number of ignitions}}{\text{number of impacts}} \times 100 \right)$$

of 50% is sometimes taken as the sensitivity figure. There is, however, considerable variation in the values quoted for the impact sensitivity of a liquid explosive such as nitroglycerine. Experiments by *Bowden, Mulcahy, Vines* and *Yoffe* (1947) using flat strikers on nitroglycerine spread as a continuous film on a flat anvil also failed to give consistent results, and as a consequence a more detailed study was made of the physical processes occurring during impact. It was found that the shape of the striker and the distribution of the explosive were most important.

### 3.1. Effect of gas bubbles on impact sensitivity of liquids

One observation which is of major importance for our present purpose is the profound effect which minute bubbles of included gas have on the impact sensitivity. It is found that, under many conditions of impact, minute gas bubbles are present or are trapped in the explosive. If one of the surfaces contains a small cavity (which may be less than a millimetre in diameter and in depth), inclusion of air almost invariably occurs during impact (see fig. 10), and even if smooth flat surfaces are used air is trapped if the distribution of the explosive is not uniform. In fig. 11 the nitroglycerine is spread as a ring, and when this is struck with a flat hammer the small amount of gas in the centre is trapped. If the nitroglycerine is spread as two parallel strips (fig. 12) or as a number of small drops (fig. 13) or if a small air bubble is blown into a continuous film of explosive (fig. 14), again small air spaces are trapped during impact (see fig. 15). If a small amount of gas is present the explosive may be detonated by the gentlest of blows. With nitroglycerine, for example, when a bubble as small as $5 \times 10^{-3}$ cm. in radius is present, an explosion efficiency of 100% may be obtained with a 40 g. weight falling 10 cm. (see Table VI).

When no gas bubble is present very high energies of the order of $10^5$ to $10^6$ g.cm. and high rates of approach are necessary for

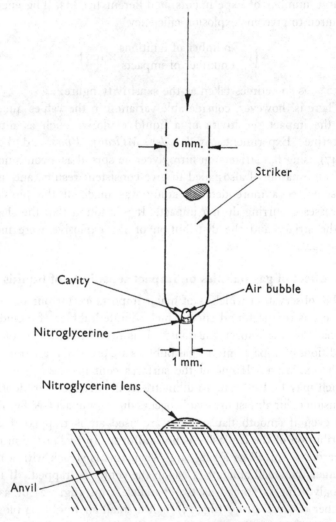

Fig. 10. Cavity striker showing small air bubble inside cavity.

explosion. With curved strikers containing a cavity explosion may occur even with a 40 g. striker falling 0·5 cm. (i.e. energy 20 g.cm.). At this energy of impact the velocity of approach of the surfaces is slow and is of the order of 30 cm.sec.$^{-1}$.

TABLE VI. *Explosion efficiency using a curved brass striker with a cavity on a brass anvil*

| Mass of striker (g.) | Height of fall (cm.) | Energy (g.cm.) | Explosion efficiency (%) |
|---|---|---|---|
| 40 | 10 | 400 | 100 |
| 40 | 5 | 200 | 80 |
| 40 | 3 | 120 | 17 |

Fig. 11. Nitroglycerine spread as a ring on a flat anvil (× 2).

Fig. 12. Nitroglycerine spread as two parallel strips on a flat anvil (× 2).

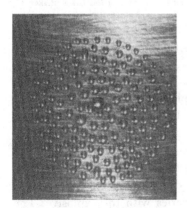

Fig. 13. Nitroglycerine spread as a number of small drops on a flat anvil (× 2).

Fig. 14. Small air bubble blown in nitroglycerine film (× 2).

When the nitroglycerine is spread as a film containing a gas bubble, as in figs. 11–14, and subjected to impact between flat surfaces, the energies required for explosion are again low and are similar to those in Table VI. A 100% efficiency may be obtained with a 60 g. striker falling about 15 cm. (energy 900 g.cm.). When no gas bubble is present no explosions are obtained at this energy of impact.

This effect of small gas spaces on the impact sensitivity is quite general and may be observed with any explosive provided it is in the liquid or plastic state, e.g. nitroglycerine, nitroglycol, diglycol dinitrate, diglycerol tetranitrate, 1.2.4.butane-triol-trinitrate,

Fig. 15.  Small air pockets trapped during impact of a flat brass striker on nitroglycerine spread as drops on a mica surface.

1.4.butane-diol-dinitrate; methyl nitrate, ethyl nitrate; nitromethane, hydrogen peroxide-methyl alcohol mixtures, tetranitromethane-toluene mixtures; molten T.N.T., picric acid, trinitroanisole, tetryl, R.D.X., P.E.T.N., blasting gelatine.

These small bubbles may easily be trapped under many conditions of laboratory and practical operation. Since they are difficult to see even with a microscope, they readily escape detection and, under appropriate conditions, they may constitute a serious hazard in the handling of explosives.

The intensity of the explosion varies with the explosive. The energy required to initiate these explosions with, for instance, a cavity striker is much the same for all. However, with some explosives such as ethyl nitrate where the oxygen balance is not favourable and as a consequence the heat of reaction is not so great, the explosions are of a local character and do not propagate into the surrounding explosive.

## 3.2. Initiation of explosion by the adiabatic compression of trapped gas bubbles

The initiation of explosion when small gas bubbles are trapped in the liquid during impact is due to the adiabatic compression and heating of the bubble. For an ideal gas the temperature reached inside the bubble ($T_2$ degrees absolute) depends on the compression ratio and is given by

$$T_2 = T_1 \left(\frac{P_2}{P_1}\right)^{(\gamma-1)/\gamma}, \qquad (3.1)$$

where $P_1$ is the initial pressure inside the bubble, $P_2$ the final pressure and $\gamma$ the ratio of the specific heats.

There is some evidence that explosion can take place when the minimum compression ratio is $c$. 20:1, and approximate calculations suggest that the minimum temperature rise in the gas bubble must be $c$. 450° C. for initiation to occur. Pressures and temperatures considerably greater than this are attained even in light impact.

(1) *Impact experiments at high initial gas pressures.* The final temperature $T_2$ will depend on the initial gas pressure $P_1$ and on $\gamma$. If explosion requires a definite high temperature $T_2$ to be reached, then initiation will be the more difficult the greater the initial pressure (*Yoffe*, 1949). This has been tested in a closed vessel in which the initial gas pressure $P_1$ could be varied between 1 and 100 atm. A sketch of the apparatus is given in fig. 16. Nitroglycerine is spread as a ring on a flat steel anvil inside a chamber which is then filled with air or nitrogen from a gas cylinder. When the gas pressure reaches a required value the explosive is hit with a flat steel hammer. Some results are given in Table VII.

For an impact energy of $5 \times 10^3$ g.cm. an explosion efficiency of 100 % is obtained when the initial gas pressure is 1 atm. When the initial gas pressure is 20–30 atm. no explosions are obtained at the same effective impact energy.

(2) *Impact experiments at low initial air pressures*. The effect of reducing the initial air pressure below atmospheric has also been tried. Assuming the final pressure $P_2$ to be constant for a given

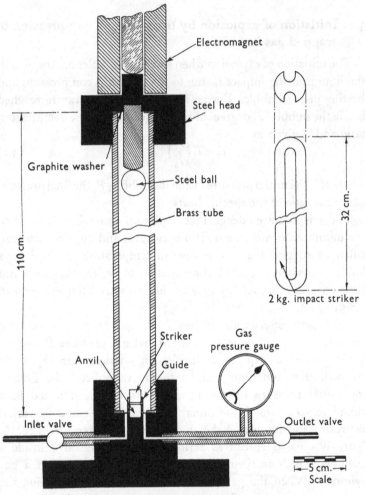

Fig. 16. Apparatus for impacts at high initial gas pressures.

impact energy the final temperature $T_2$ increases as the initial pressure $P_1$ is reduced. Under these conditions, however, the mass of gas trapped in the bubble becomes smaller and the quantity of heat developed is correspondingly lowered.

TABLE VII. *Explosion efficiencies obtained with a ring of nitro-glycerine at different initial pressures $P_1$ of air and nitrogen*

| Mass of striker (g.) | Height of fall (cm.) | Kinetic energy of impact (g.cm.) | Initial gas pressure $P_1$ (atm.) | Explosion efficiency | |
|---|---|---|---|---|---|
| | | | | Nitrogen (%) | Air (%) |
| 112 | 30 | $3\cdot4 \times 10^3$ | 1 | 83 | 90 |
| | 36 | | 10 | 30 | — |
| | 36 | | 20 | 0 | — |
| | 36 | | 25 | — | 20 |
| | 36 | $3\cdot4 \times 10^3$ | — | — | 0 |
| 112 | 63·5 | $7\cdot1 \times 10^3$ | 1 | 100 | 100 |
| | 98 | | 20 | 10 | — |
| | 98 | | 25 | 0 | — |
| | 98 | | 30 | — | 36 |
| | 98 | $7\cdot1 \times 10^3$ | 35 | — | 0 |

The apparatus used is shown in fig. 17 and consists of a collapsible bellows which is so designed that changes in gas pressure do not cause any relative movement of striker and anvil. The nitroglycerine is spread as a ring on the anvil.

It is found that the explosion efficiency for low-energy impacts (300–500 g.cm.) is appreciable even when the initial air pressure is less than $10^{-5}$ mm. Although the other gases have been pumped out the vapour pressure of nitroglycerine itself at room temperature is $c$. $10^{-3}$ mm. The experiments suggest that the initiation of explosion under these conditions is due to the compression and adiabatic heating of the nitroglycerine vapour itself. This implies that the vapour is not condensed during the rapid compression produced by the impact.

(3) *Impact experiments in low-pressure atmospheres of different gases.* When gases having a lower $\gamma$ are included there is a fall in explosion efficiency (cf. equation (3.1)). With these gases the final temperature $T_2$ is lower than that obtained when the trapped gas is air. Values have been obtained (in the apparatus shown in fig. 17) for the explosion efficiency over a fairly wide range of initial gas pressures.

Some results are given in figs. 18 and 19 where there is a comparison of the explosion efficiency when air $\gamma = 1\cdot4$, normal pentane $\gamma = 1\cdot08$, and ethylene $\gamma = 1\cdot26$ are used. Both pentane and ethylene are less efficient than air for which the efficiency remains at 100 % even down to pressures of $10^{-5}$ mm. (fig. 19). With methyl nitrate vapour the explosion efficiency remains unaltered when the evacuated space is filled to a pressure of 50 mm. with the vapour.

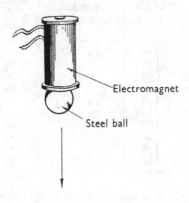

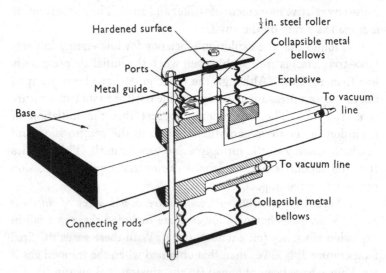

Fig. 17. Low-pressure apparatus (section through centre).

These experiments also demonstrate the sensitizing effect of bubbles of normal pentane, carbon tetrachloride and methyl nitrate. These are gases which normally condense as the pressure is increased. In the impact experiments, however, the compression occurs very rapidly (*c.* $10^{-4}$ sec.), and it appears that under these conditions condensation does not occur. The fact that the explosion efficiency with these gases is less than for air supports the view that initiation is due to the adiabatic compression of trapped gas.

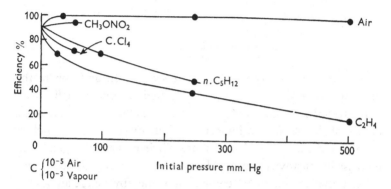

Fig. 18. Explosion efficiency for different initial pressures of air, normal pentane, ethylene, carbon tetrachloride and methyl nitrate. Mass of striker 95 g., height of fall 30 cm.

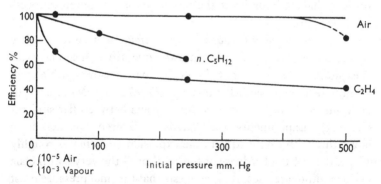

Fig. 19. As for fig. 18. Mass of striker 210 g.

With cavity strikers (fig. 10) a clear-cut distinction was obtained by *Mulcahy* (1948) when gases having a low value for $\gamma$ were substituted for air. The gases used were ether $\gamma = 1 \cdot 08$ and carbon tetrachloride $\gamma = 1 \cdot 13$. Some results are given in Table VIII. It will be seen that no explosions are obtained with carbon tetrachloride or ether vapour.

TABLE VIII. *Cavity impact in gases of different* $\gamma$

Mass of striker 205 g. Height of fall 24 cm. Initial gas pressure $p_1 = 6\cdot4$ cm.

| Atmosphere | Explosion efficiency (%) |
|---|---|
| Air | 78 |
| Ether | 0 |
| Carbon tetrachloride | 0 |

(4) *Oxidizing properties of the trapped gas.* In addition to the physical properties (e.g. thermal conductivity and specific heat) of the trapped gas influencing the explosion efficiency the chemical nature of the gas is also important. Thus oxidizing gases such as oxygen and nitrous oxide are more efficient than nitrogen. This increase in sensitivity is probably due to a reaction with the oxygen of the initial products of decomposition whereby a large amount of heat is liberated. The high explosion efficiency observed when the trapped gas is methyl nitrate may be attributed to the exothermic character of the decomposition of methyl nitrate. It is clear, however, that the initiation in the vapour phase is a complex process (see later § 2.9).

(5) *Pressure developed in gas before explosion.* In order to calculate from equation (3.1) the temperature and quantity of heat necessary for explosion, it is necessary to know the final pressure $P_2$ which will just initiate an explosion. Estimates by *Eirich* and *Tabor* (1948) put this value at about 20–100 atm. for impacts between flat surfaces (see § 3.3). Experiments using anvils of different hardness have shown that with cavity strikers the explosion efficiency is roughly independent of the hardness of the anvil until the very soft metals such as indium are used. This suggests that the final pressure must be *c.* 400 atm. for cavity strikers.

A direct measurement of the transient pressure during impact has been made by *Gray* (1949) using the piezoelectric effect. The apparatus used is shown in fig. 20 and was designed so as to simulate conditions obtaining in cavity strikers. Suitable electronic apparatus makes it possible to obtain graphs of pressure against time. With non-explosive liquids the curve obtained has a single peak (see fig. 21). In fig. 22 the pressure rises to a maximum value of 250 atm.

and falls to zero in a time of the order of 250 $\mu$sec. The results obtained demonstrate that even in light blows quite high pressures are reached. When the apparatus contains nitroglycerine and a small bubble of air ($10^{-4}$ ml.) is trapped, the result shown in fig. 23 is obtained. When a sufficient pressure is reached detonation of the

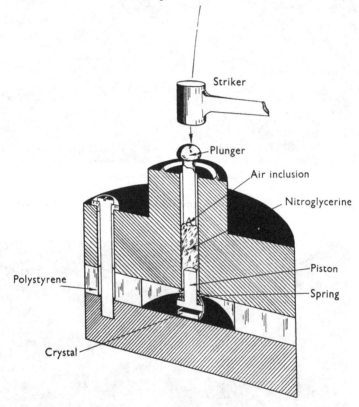

Fig. 20. Apparatus for measuring pressure developed in a liquid during impact.

nitroglycerine occurs, and the added thrust on the quartz crystal gives rise to a change in the slope of the curve. The point $A$ corresponds to metallic contact between hammer and the plunger. $B$ is the point at which the pressure in the liquid begins to rise appreciably. Initiation of explosion follows at $C$, 150 $\mu$sec. after $B$ and at a pressure of 610 atm.

(6) *Temperature rise for explosion.* The pressures obtained in the impact apparatus above imply very high transient temperatures,

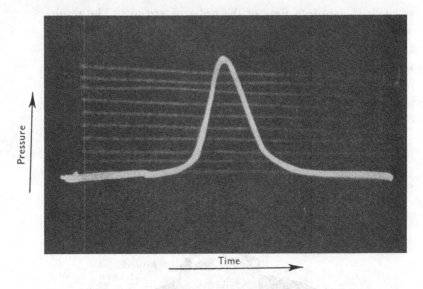

Fig. 21. Variation of pressure with time during light impact on nitroglycerine.

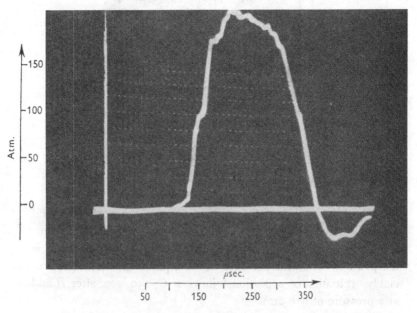

Fig. 22. Pressures during impact on a non-explosive liquid. A 75 g. hammer falling 7 cm. on to $\frac{1}{30}$ ml. of liquid in the presence of a small air bubble.

and even though part of the heat is lost by conduction, the gas bubble, and hence the explosive vapour in it, will easily rise in temperature by 1000° C.

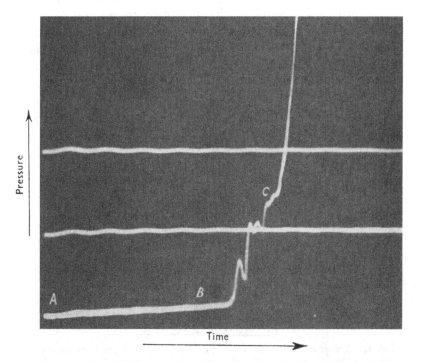

Fig. 23. As for fig. 22 using nitroglycerine. Energy of fall 280 g.cm. Volume of nitroglycerine 0·006 ml. Explosion occurred at $C$ (610 atm.), 150 $\mu$ sec. after the initial rise of pressure.

An estimate of the temperature necessary for explosion has also been made, using the apparatus shown in fig. 24 (*Bowden* and *Gurton*, 1949). This consists of a piston and cylinder, and the air in the cylinder is compressed between known volume limits by means of a weight falling on the piston. The explosive is not itself subjected to impact, but is ignited by the hot gas produced during the compression. The final temperature $T_2$ is given by

$$T_2 = T_1 \left(\frac{v_1}{v_2}\right)^{\gamma-1} \tag{3.2}$$

A volume compression ratio of *c.* 20:1 is required to initiate the explosion of nitroglycerine, and this corresponds to a temperature of 450–480° C.

Since the mass of gas trapped in a bubble during the impact experiments is known, it is possible to estimate the quantity of heat developed in the trapped gas during impact. This is usually of the order of $10^{-7}$ to $10^{-10}$ cal. for small gas bubbles whose initial diameter is *c.* 1·0 mm.

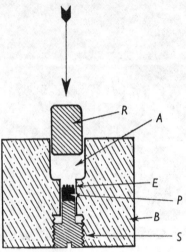

Fig. 24. Apparatus for the initiation of explosion by the rapid compression of air showing explosive E, piston R, block B, screw S, piston P, air cavity A.

(7) *Time from impact to initiation.* Because the initiation of explosion by impact is thermal in origin the magnitude of the temperature that is required to initiate an explosion will depend on the duration of the temperature flash. A determination of the time interval between the first moment of impact and the onset of explosion has been made both for cavity strikers (fig. 10) and for explosion initiated between flat surfaces. An optical method in conjunction with a rotating drum camera was used.

*Initiation using cavity strikers.* A sketch of the apparatus used for the initiation of explosion by cavity strikers is given in fig. 25. A cavity striker is allowed to fall into a lens of nitroglycerine placed directly over the slit which was made of a semi-transparent

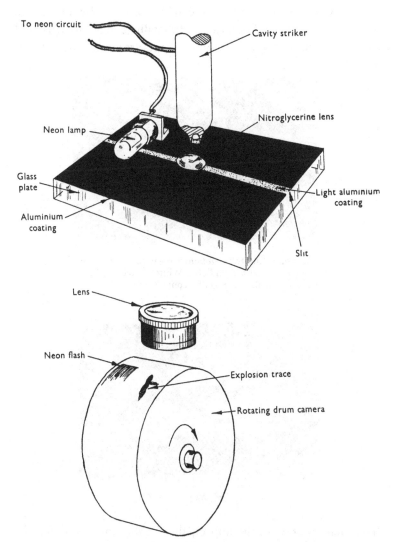

Fig. 25. Apparatus for measuring the time from impact to
explosion for cavity strikers.

aluminium coating that had been evaporated on to a glass slab. The slit is set at right angles to the direction of rotation of a drum camera. When explosion occurs the light is recorded by the camera. At the same time the first instant of metallic contact is registered by a flash from the neon lamp. Some results are given in figs. 26–28.

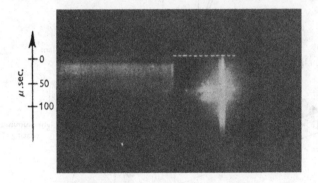

Fig. 26. Light trace obtained from cavity explosion (right) and neon flash (left). White dotted line indicates the instant of metallic contact.

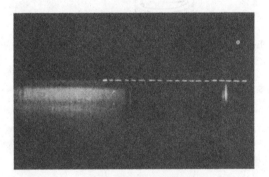

Fig. 27. As for fig. 26.

The total duration of the impact, that is, the time for which the striker is in contact with the anvil is of the order of 200–300 $\mu$sec. In many experiments it is found that explosion begins within a few microseconds of the first moment of metallic contact.

*Initiation between flat surfaces.* The apparatus used in this experiment is shown in fig. 29. Here nitroglycerine is spread as two parallel strips along the slit (fig. 30). A beam of light from the lamp *L* is passed vertically through the slit and the strips are struck with a cylindrical glass hammer *H*. During impact the strips flow together and trap a small amount of air. Coalescence is marked by extinction of the light passing through the slit. Fig. 31*a* shows one experiment when no explosion took place, and *AA'* marks the coalescence of the strips, i.e. the moment at which rapid compression of the trapped air occurs. In figs. 31*b* and *c* explosion has taken place at *B*. The average value for the time between rapid compression and the onset of explosion is 20 $\mu$sec.

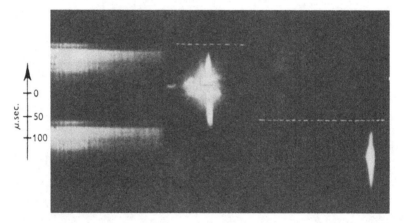

Fig. 28. As for fig. 26.

Such short time delays from impact to initiation (*c*. $10^{-5}$ sec.) indicates that the temperature of the bubble must be high, and this is in accord with the measurements made in the previous section.

(8) *Non-ideal behaviour of gases during compression.* Calculation of the final temperature $T_2$ by means of equation (3.1) for vapours such as pentane give values that are well below the expected ignition temperature for nitroglycerine. One possible explanation for this lies in the non-ideal behaviour of the vapours used.

For an ideal gas, where

$$pV = RT,$$

the relation between temperature and volume for an adiabatic compression is given by the equation (3.2), viz.

$$TV^{\gamma-1} = \text{const.},$$

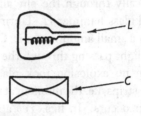

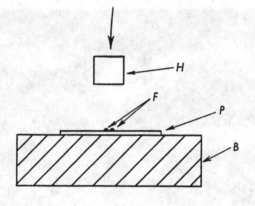

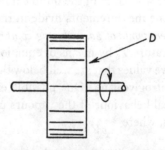

Fig. 29. Arrangement of optical apparatus for strip experiments with drum camera showing lamp *L*, condensing lens *C*, transparent glass striker *H*, strips of nitroglycerine *F*, glass plate *P*, steel anvil with slit (not shown) *B*, camera lens *K*, and rotating drum *D*.

and this $\gamma$ is
$$\gamma = \frac{C_p}{C_v} = 1 + \frac{R}{C_v}$$
$$= 1 + \frac{R}{n}, \tag{3.3}$$

where $n$ is the number of 'square terms' in the expression for the internal energy.

Lewis and von Elbe (1939) have shown that $\gamma$ may depend on the rate of compression and suggest that $\gamma$ will rise during the compression of polyatomic molecules because of the failure of the kinetic

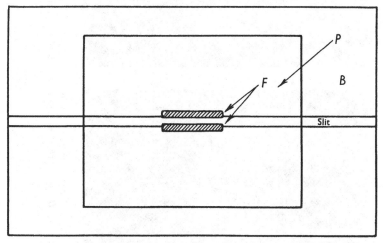

Fig. 30. Strips of nitroglycerine $F$ placed on a glass plate $P$ so that they lie parallel to slit.

energy to share itself out among all the modes of motion. A reduction in the number of degrees of freedom $n$ will result in an increase in $\gamma$. However, the compression time before explosion is fairly long (c. $10^{-5}$ sec.) compared with the time taken for energy to be shared among the various modes of motion (Alexander and Lambert, 1942). It is unlikely, therefore, that Lewis and von Elbe's theory can be used to modify equation (3.1).

In the initiation of explosion by the compression of gas spaces, the pressure ratio is high, and deviations from the ideal nature of the gas must be taken into account. If van der Waals's relation

$$\left(P + \frac{a}{V^2}\right)(V - b) = RT$$

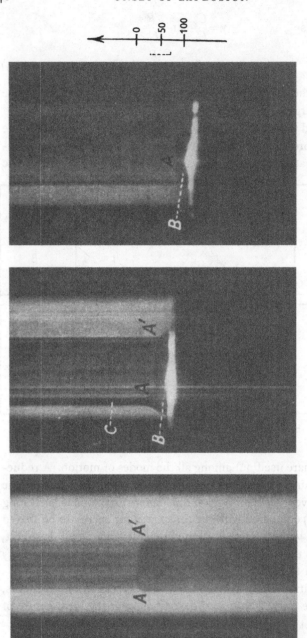

0   50   100

Fig. 31a. Coalescence of strips arranged as in fig. 28. No explosion occurred. Edge of shadow AA' represents the instant of coalescence of strips.

Fig. 31b. As for fig. 31a but with explosion at B some 16 μsec. after coalescence.

Fig. 31c. As for fig. 31a with explosion at B very shortly after coalescence at A.

is used instead of the ideal equation of state the relation between temperature and volume for an adiabatic compression becomes

$$T(V-b)^{\gamma-1} = \text{constant}$$
$$= K, \qquad (3.4)$$

where the value of $\gamma$ that may be used is

$$\gamma = 1 + \frac{R}{C_v}.$$

This gives a relation between temperature and pressure

$$P = \frac{R}{(K)^{1/(\gamma-1)}} T^{\gamma/(\gamma-1)} - \frac{a}{\left\{\left(\frac{K}{T}\right)^{1/(\gamma-1)} + b\right\}^2}. \qquad (3.5)$$

The values for the final temperature calculated from equation (3.5) are more in accord with experiment than those calculated from (3.1). This provides a possible explanation for the sensitizing action of the condensable gases and of the self-sensitized explosion of nitroglycerine spread as a ring *in vacuo* ( $< 10^{-5}$ mm. Hg). Here it is the vapour of the nitroglycerine itself which undergoes the adiabatic heating and which initiates the explosion (see § 3.2 (2)).

(9) *Explosion initiated in the vapour phase.* It is suggested that the initiation begins in the vapour phase. That is to say the ignition starts as a burning of the explosive vapour inside the hot bubble and spreads to the liquid. One apparent difficulty, however, is the low concentration of explosive vapour in the gas bubble. The vapour pressure of nitroglycerine ($10^{-3}$ mm. at room temperature) corresponds in an air bubble at atmospheric pressure to a millionfold dilution. Because of the rapidity of the compression and its magnitude, the relative concentration of explosive vapour probably cannot increase very much during the sudden heating. The inflammation of such very dilute mixtures with inert gases of nitroglycerine and of the analogous explosives methyl nitrate and ethyl nitrate has been examined (*Gray* and *Yoffe*, 1949*a*, *b*). It is found that all these nitrates when diluted from a thousand to a million times with argon or nitrogen, glow on admission to a hot quartz vessel (see Table IX). This blue glow has been examined over a temperature range of 300–500° C. The total pressure rarely exceeds a few cm. of mercury. It is clear that ignition of these explosive

vapours can occur when their concentration is remarkably low, and this behaviour supports the view that a vapour phase inflammation precedes decomposition in the liquid.

TABLE IX. *Total pressure (mm.) at which a mixture of explosive vapour and diluent glows on admission to quartz vessel*

Temperature approx. 420° C.

| Relative concentration of explosive vapour | Explosive vapour used | | |
|---|---|---|---|
| | Methyl nitrate | Ethyl nitrate | Nitro-glycerine |
| Pure vapour | 0·063 | 0·035 | — |
| 1 in 100 | 0·5 (N₂) | 0·2 (A) | — |
| 1 in 1,000 | 1·0 (N₂) | 0·4 (A) | — |
| 1 in 10,000 | 5 (A) | 1·0 (A) | 5 (A) |
| 1 in 100,000 | 50 (A) | — | 50 (A) |
| 1 in 1,000,000 | ~60 (N₂) at 470° C. | — | — |

Symbols in parentheses denote diluent used: A, argon; N₂, nitrogen.

The glow that is observed with alkyl nitrate vapours resembles in appearance the cool flame obtained during the oxidation of certain hydrocarbons, and it is thought that it is connected with the formation during decomposition of excited formaldehyde molecules. With pure methyl nitrate the glow occurs when the pressure is a fraction of a millimetre. As the pressure of methyl nitrate vapour is increased to several millimetres a violent explosion may follow the glow. The relation of the glow and explosion regions are shown in fig. 32.

In addition to facilitating the glow of alkyl nitrates, addition of inert gases also assists explosion. The experimental results suggest that the explosion of the alkyl nitrate vapours such as methyl nitrate proceeds by a 'chain thermal' mechanism (*Gray* and *Yoffe*, 1950).

The initial step in the decomposition of the vapour is the formation of a free radical and nitrogen dioxide. Phillips (1947, 1950) has suggested the following mechanism for the initial stages

$$RCH_2ONO_2 \rightleftharpoons RCH_2O^- + NO_2, \tag{1}$$

$$RCH_2O^- + RCH_2ONO_2 \rightarrow RCH_2OH + R\overset{|}{C}HONO_2, \tag{2}$$

$$R\overset{|}{C}HONO_2 \rightarrow RCHO + NO_2, \tag{3}$$

or $$RCH_2O^- \rightarrow \tfrac{1}{2}RCHO + \tfrac{1}{2}RCH_2OH, \tag{4}$$

followed by oxidation reduction reactions between the products to give gases such as NO and CO.

When liquid explosives are detonated by light impact, the following sequence of events is therefore postulated. Sudden compression and heating of the trapped gas occurs and exothermic

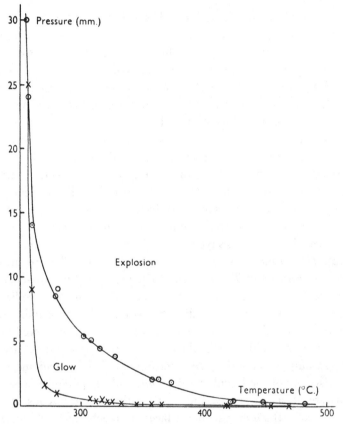

Fig. 32. Inflammation of methyl nitrate showing glow and explosion regions.

decomposition of the explosive vapour begins. Even in the absence of air the vapour of the explosive itself suffers on compression a considerable temperature rise. The physical and chemical heating thus induced leads to further evaporation of explosive from the walls of the bubble to give a richer mixture than existed before impact. Inflammation of the explosive vapour near the walls may then become sufficiently violent to ignite the liquid itself, and the explosion grows as a rapid burning which may spread through the liquid (see Chapter V).

### 3.3. Initiation of explosion in liquids in the absence of gas spaces

(1) *Heating due to viscous flow.* If special precautions are taken to exclude the possibility of trapping small gas spaces in a liquid during impact then very much higher energies are necessary to initiate explosions. With flat surfaces, impact energies of the order of $10^5$–$10^6$ g.cm. are required when gas is carefully excluded. This should be contrasted with values of 20–100 g.cm. which can cause initiation if a gas bubble is trapped.

*Cherry* (1945) and *Eirich* and *Tabor* (1948) have calculated that at these high energies, temperatures of the order of several hundred degrees may be obtained near the periphery of the striker, owing to the viscous heating of the liquid as it escapes rapidly between the impacting surfaces. It is suggested that in the absence of gas spaces the initiation under extreme conditions of impact may be due to a localized heating of this type.

(2) *Temperature rise for low-energy impacts.* The treatment given by *Eirich* and *Tabor* (1948) holds fairly well for low-energy impacts on liquid films. They consider the case of a collision between a flat hammer of radius $R$ and mass $M$ approaching the anvil with an initial velocity $V_0$ (see fig. 33). The anvil is covered with a thin film of liquid of initial thickness $h_0$, viscosity $\eta$ and density $\rho$. The assumptions are that the surfaces are infinitely rigid and that the viscosity of the liquid is independent of temperature pressure and rate of shear. *Eirich* and *Tabor* estimate the viscous forces developed in the liquid as the surfaces approach. If at any instant the hammer velocity is $V$, the film thickness $h$, the pressure in the film $p$, and $c$ is the radial velocity of flow of the liquid in any plane $z$, then the basic viscous equation, ignoring inertia effects, is

$$\frac{\partial^2 c}{\partial z^2} = \frac{1}{\eta}\frac{dp}{dr}. \tag{3.6}$$

Hence

$$c = \frac{1}{2\eta}\frac{dp}{dr}\left(z^2 - \frac{h^2}{4}\right). \tag{3.7}$$

The velocity profile is thus parabolic.

The volume of liquid squeezed out through the boundary at any instant is equal to the volume swept out by the hammer in the same instant. Simple geometry shows that the mean radial rate of flow $\bar{C}$ of the liquid at any instant is $\bar{C} = \dfrac{rV}{2h}$. Using this relation, integration of equation (3.7) yields

$$p = \frac{3\eta V(R^2 - r^2)}{h^3},\qquad(3.8)$$

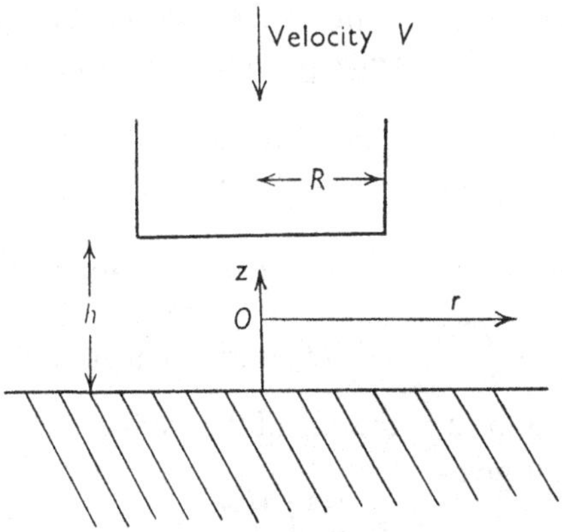

Fig. 33. Collision of flat parallel surfaces separated by a liquid layer of thickness $h$.

so that the pressure is a maximum at the centre of the hammer and falls off parabolically towards the periphery. The maximum value is given by

$$p_{\text{max.}} = \frac{3\eta V R^2}{h^3}.\qquad(3.9)$$

The velocity of the hammer may be determined at any stage of the collision if, in any short interval of time, the kinetic energy lost by the hammer, $MV\,dV$, is equal to the work done on the liquid in that interval, $dh \int 2\pi r p\,dr$. The solution of the differential equation yields

$$V = V_0 - \frac{3\pi\eta R^4}{4M}\left(\frac{1}{h^2} - \frac{1}{h_0^2}\right).\qquad(3.10)$$

By inserting this value of $V$ in equation (3.8) the pressure in the liquid film may be calculated as the collision proceeds. The pressure increases rapidly as the liquid is squeezed out, reaches a maximum and falls rapidly to zero as the striker is brought to rest. The velocity of extrusion of the liquid film and the rate of shear developed in the liquid film, rise to a maximum and fall off to zero in a similar way during the course of the collision.

The energy used in overcoming the viscous forces in the liquid film is dissipated as heat. The maximum effect occurs at the solid surfaces $z = \pm \frac{1}{2}h$ at $r = R$, since here the rate of shear is a maximum and, as the radial velocity is zero, the element of the liquid remains in shear for the whole of the interval of impact. The rate of dissipation of energy $H$ by viscous flow per unit volume of liquid per second is given by

$$\frac{dH}{dt} = \eta \left(\frac{dc}{dz}\right)^2, \qquad (3.11)$$

substituting for the value of $\frac{dc}{dz}$ from equation (3.7), replacing $\frac{dH}{dt}$ by $\frac{dH}{dh}\frac{dh}{dt} = -\frac{dH}{dh}V$, and integrating equation (3.11) from the initial instant of impact to the end of the collision when the hammer is brought to rest gives

$$H = \frac{0.31}{R^4} \left(\frac{M^3 V_0^5}{\eta}\right)^{\frac{1}{4}}. \qquad (3.12)$$

Neglecting heat lost in the liquid by convection and by conduction to the solid, the temperature rise $T°$ C. in the liquid near the surface may be calculated. Some results for the maximum pressure $p_{max.}$, radial velocity of flow and maximum rate of shear for a liquid similar to nitroglycerine are given in Table X. The values used are $\eta = 25$ centipoise, specific gravity $= 1.6$, specific heat $= 1.8$ cal.g.$^{-1}$ $°$ C.$^{-1}$.

TABLE X

| Mass of hammer (g.) | Velocity of hammer prior to impact (cm.sec.$^{-1}$) | $p_{max.}$ (atm.) | $\bar{C}$ (metres per sec.) | $S_{max.}$ (cm.sec.$^{-1}$ cm.$^{-1}$) | $T$ ($°$ C.) |
|---|---|---|---|---|---|
| 40 | 150 | 21 | 30 | $1.2 \times 10^7$ | 2 |
| 400 | 200 | 1360 | 142 | $2.0 \times 10^7$ | 98 |

It will be seen that for light impacts high pressures are easily reached but the temperature rise is not great. For the two impact energies given above it is 2 and 100°C. respectively. *Cherry* has shown that if allowance is made for conduction, these temperatures are reduced by about one-half. These temperatures are well below the ignition temperature so that viscous heating plays little part in the initiation of explosion by gentle impact.

(3) *Temperature rise for high-energy impacts.* When high-energy impacts are used, e.g. $10^5$–$10^6$ g.cm., *Cherry* (1945) has shown that allowance must be made for the elastic deformation of the solid surfaces. As a result of this deformation there is a pool of liquid below the centre of the hammer and the maximum pressure is much lower than that calculated by *Eirich* and *Tabor*. Nevertheless, the temperature rise can still be several hundred degrees.

Even the analysis by *Cherry* does not take into account the variation of viscosity with temperature and pressure, so that it can not be used for accurate calculations of the temperature rise. It does show, however, in a qualitative way that high temperatures can be reached during the viscous flow of a liquid if the impact energies are very high.

CHAPTER IV

# INITIATION BY IMPACT OF EXPLOSION
# IN SOLIDS

With solid explosives we have an additional possible source of hot spots, namely friction between the crystal particles themselves. Since the maximum temperature which can be reached is limited by the melting-point of the solid, we may note an interesting difference between the primary and secondary explosives. Most of the primary explosives will detonate below their melting-points so that intercrystalline friction may initiate the explosion. The majority of secondary explosives however, melt at a temperature below that at which rapid decomposition sets in so that inter-granular friction will not in general cause initiation. There is evidence (*Bowden et al.* 1947; *Yoffe*, 1949) that with the solid secondary explosives the compression of the small gas spaces which are present between the crystals can again be responsible for initiation. Under impact, plastic flow of the solid occurs so that the gas is sealed in and compressed. The general behaviour is analogous to that of liquids, but the impact energies to produce the sealing off and compression of the gas are appreciably higher than with liquids. It is also shown that the action of grit in solid explosives under impact is similar to the behaviour under friction, in that initiation is determined by the melting-point of the grit. It is only under very severe conditions of impact that we could expect ignition to occur by viscous heating of the flowing solid (see, however, Rideal and Robertson, 1948*a*).

## 4.1. Initiation by the adiabatic compression of trapped gas

(1) *Distribution of the explosive.* When a solid such as P.E.T.N. is spread on an anvil as a ring rather than as a uniform film of crystals, the sensitivity to impact is increased (see fig. 34): the energy necessary to give an explosion efficiency of 50% is $2 \cdot 8 \times 10^4$ g.cm. for a ring, and $7 \cdot 1 \times 10^4$ g.cm. for a continuous film. This increase in sensitivity when the explosive is spread as a ring is

analogous to that of liquid explosives, although the effect is not so marked.

(2) *Impact experiments at high initial gas pressures.* The experiments with nitroglycerine at a high initial gas pressure confirmed the view that the initiation of explosion in liquids by gentle impact is due to the adiabatic compression of trapped gas. Similar experiments with P.E.T.N. confirm the observation that the

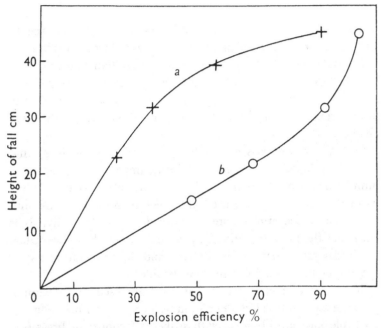

Fig. 34. Explosion efficiency against height of fall curves for P.E.T.N. spread as a continuous film *a*, and as a ring *b*. Mass of striker 1860 g.

presence of trapped gas has a sensitizing effect and they show that the explosion efficiency depends on the initial gas pressure $P_1$. The apparatus used was that shown in fig. 16 except that the hardened steel weight weighing 2 kg. was employed instead of the spherical ball.

Some results for P.E.T.N. spread as a ring are given in Table XI. Measurements of the time of fall of the weight *O* showed that increase in the viscous resistance of the air produced no appreciable reduction in velocity.

TABLE XI. *Explosion efficiency obtained with a ring of*
*P.E.T.N. at a high initial pressure of nitrogen*

| Height of fall (cm.) | Initial gas pressure (nitrogen) ($P_1$ atm.) | Kinetic energy of impact (g.cm.) | Explosion efficiency (%) |
|---|---|---|---|
| 23 | 1 | $4·6 \times 10^4$ | 58 |
| 25 | 1 | $5 \times 10^4$ | 100 |
| 36·5 | 100 | $5 \times 10^4$ | 13 |

There is a reduction in the explosion efficiency from 100 to
13% when the initial gas pressure is increased from 1 to 100 atm.
This result provides strong evidence that the initiation of explosion
in P.E.T.N. is due to the sudden compression of trapped gas
spaces. An initial high gas pressure also has an effect on the
propagation of explosions. This will be discussed in the next
chapter.

(3) *Effect of hardness of the striker.* The final pressure $P_2$ that
can be developed in the solid film during impact is very high, the
limit being set by the dynamic flow pressure of the striker. When
hardened steel strikers are used, the pressure in the P.E.T.N. film
can reach 80,000 atm. before the steel deforms plastically. It is
clear that the pressure ratio $P_2/P_1$ can still be very high even when
the initial gas pressure $P_1$ is 100 atm. and the temperature rise in
any trapped gas space can be considerable.

If strikers of metals softer than steel are used the maximum
pressure attainable is less. For an impact energy of $1·6 \times 10^5$ g.cm.
no explosions are obtained with strikers of copper or brass (see
Table XII). During the impact both the copper and brass strikers
are deformed by the crystals of P.E.T.N. and a deep imprint of the
explosive, which is spread as a ring, is left on the face of the striker.
The ring of P.E.T.N. appears not to flow at all and its dimensions
are roughly the same before and after impact. When mild steel
strikers are used, explosions are readily obtained. Examination of
the surface of the steel striker after explosion shows that the
explosive has flowed, and the steel itself shows only slight plastic
deformation. Thus in order to initiate explosion it is necessary to
make the P.E.T.N. flow, and for this to occur the hardness of the
striker must be *c.* 200 kg./mm.². This means that the P.E.T.N.

TABLE XII. *Explosion efficiencies. The sensitiveness to shock of P.E.T.N. spread as a ring on a hard steel anvil under different initial gas pressures depends on the hardness of the striker*

Mass of striker 2 kg.

| Height of fall (cm.) | Striker | Hardness of striker Vickers (kg./mm.$^2$) | Initial gas pressure nitrogen ($P_1$ atm.) | Explosion efficiency |
|---|---|---|---|---|
| 46 | Mild steel | 200 | 1 | 6/6 |
| 82 | Copper | 72 | 1 | 0/5 |
| | Brass | 120 | 1 | 0/3 |
| | Mild steel | 200 | 1 | 6/6 |
| | Mild steel | 200 | 100 | 2/10* |

\* (These two explosions were of a very localized character.)

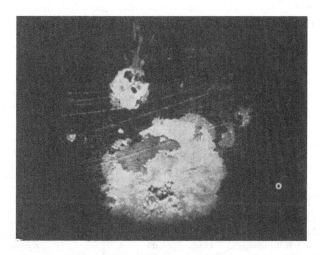

Fig. 35. Melting of sulphur during impact between flat surfaces. Striker 1860 g. falling 45 cm.

must be subjected to pressures of the order of 20,000 atm. This method of estimating pressures has been used by Taylor and Weale (1938).

(4) *Flow of explosives during impact.* When an explosive film of P.E.T.N. is subjected to impacts of high energy (*c.* $10^5$ g.cm.) a fused or flowed layer may be seen on the anvil at the edge of the continuous film. With low-melting materials such as sulphur the fused layer is even more noticeable (see fig. 35). At these impact

energies sufficient heat is not available to melt the whole of the film of P.E.T.N. The formation of the fused layer is due either to the frictional heat developed as the crystals rub over each other and the impacting surface, or to the application of a non-uniform pressure to the solid. Once the molten or plastic layer is formed, small air spaces are sealed off and the compression of these air spaces produces hot spots under impact. In other words, the melted and flowing explosive behaves just like a liquid explosive and we may expect the mechanism of initiation to be similar to that operating in liquids.

Many of the experiments described above were made on P.E.T.N. spread as a ring so that an air space was deliberately included during impact. Even when the crystals are spread as a 'continuous' film, small gas spaces are included and bubbles trapped, although they have not been deliberately added.

TABLE XIII. *Ignition temperatures*

| Explosive | Thermal ignition temp. (° C.) | Induction period (sec.) | Hot-spot temperatures during friction and impact | |
|---|---|---|---|---|
| | | | Initiation by friction (° C.) | Ignition by adiabatic compression of air (° C.) |
| Nitroglycerine | 200 | — | 450–480 | 450–480 |
| P.E.T.N. | 215 | — | 400–430 | 460–500 |
| Lead styphnate | 250 | 90 | 430–500 | 570–600 |
| Lead azide | 335 | 10 | 430–500 | 570–600 |
| Mercury fulminate | 145 | 400 | — | 630–690 |
| Tetrazene | 160 | 5 | — | 400–450 |

(5) *Hot-spot temperature.* The minimum compression ratios which must be applied to a volume of gas surrounding the explosive in order to cause initiation has been determined in the apparatus shown in fig. 24. From these results the ignition temperatures were calculated from equation (3.2) and these are given in Table XIII. The ignition temperatures obtained by this method are compared with the values obtained for hot-spot temperatures necessary for the initiation of explosion by friction

in the presence of grit. Nitroglycerine is also included for comparison.

There is a fairly close agreement between the hot-spot temperatures for initiation by friction or by adiabatic compression of air. These lie in the region of 400–600° C. It will again be noted that these momentary temperatures are appreciably higher than the standard thermal ignition temperatures.

(6) *Initiation of cyclonite and tetryl by impact.* Experiments with cyclonite and tetryl give results which follow the pattern described for P.E.T.N. though the difference in explosion efficiency of cyclonite when spread as a ring and as a continuous film is not so marked. The propagating properties of cyclonite and tetryl under impact conditions are poor. Some experiments by *Blackwood* (1950) suggest that even with black powder the initiation of explosion by impact is due to the adiabatic compression of trapped air during impact.

(7) *Impact experiments at low initial gas pressures.* The sensitivity to impact of P.E.T.N. and cyclonite when spread as a ring or as a continuous film of crystals is high even when the initial air pressure is $c.$ $10^{-5}$ mm. The vapour pressure of P.E.T.N. and cyclonite at room temperature is $c.$ $10^{-4}$ mm., and it is suggested that at low air pressures the initiation of explosion is due to the compression and adiabatic heating of the explosive vapour itself.

## 4.2. Time from impact to initiation

The time between the first application of pressure on the solid film of explosive and the initiation of explosion (as determined by the first appearance of light) was measured by means of a rotating drum camera (*Bowden* and *Gurton*, 1949). The explosive is spread on a glass plate over a slit and a flat steel roller rests on the film. The impact is produced by allowing a steel ball to fall on the roller. As soon as contact is made between the falling ball and roller, a condenser is discharged and the resulting spark is photographed on the moving film. When explosion occurs the light from it is also recorded on the same moving film thus giving the time lag between the first moment of impact and the onset of explosion. Some results obtained with various arrangements of the explosive are given in Table XIV.

A picture obtained for P.E.T.N. is shown in fig. 36 and one for cyclonite in fig. 37.

The time from impact to initiation varies from 60 to 145 $\mu$sec. according to the height of fall used. An appreciable proportion of this time will be taken up in compressing the solid film of explosive, so that the main conclusion to be drawn is that the time lag between

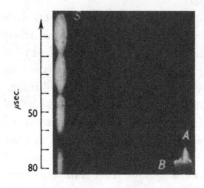

Fig. 36. Delay between impact and explosion of P.E.T.N. spread as a ring. The beginning of the spark trace $S$ represents the first instant of impact. The first appearance of the explosion light occurs at the gas pocket $A$. The impact was provided by a 530 g. ball falling 155 cm.

Fig. 37. As for fig. 36, using cyclonite.

compression and initiation is probably less than $10^{-4}$ sec. The presence of grit in the explosive film did not have much effect on the time lag. The delays observed with other explosives such as cyclonite tetryl and Dina are very similar to those quoted in Table XIV.

Rideal and Robertson (1948 a) have also estimated this time from impact to initiation. They used a different arrangement and

TABLE XIV. *Time from impact to explosion*

| Condition of explosive | Mass of striker (g.) | Height of fall (cm.) | Average delay ($\mu$sec.) |
|---|---|---|---|
| P.E.T.N. spread as a ring | 1860 | 60 | 131 |
| P.E.T.N. spread as a ring | 530 | 155 | 65 |
| P.E.T.N. spread as a continuous layer | 530 | 155 | 61 |
| P.E.T.N. spread as a continuous layer | 250 | 180 | 104 |
| P.E.T.N. containing glass particles | 250 | 180 | 106 |

measured the time interval from impact to the arrival of the explosion flame at two electrodes placed a short distance from the explosive. The time lag they observed was of the same order of magnitude but it was appreciably longer than the above. This is to be expected since it includes the time taken for the flame to travel through the explosive and to reach the electrodes (see Chapter V).

## 4.3. The influence of grit

Some data for the effect of added grit on the impact sensitivity of solid explosives have for convenience been given in Table II, Chapter II. Again it is found that the melting-point of the grit is the determining factor. With a secondary explosive such as P.E.T.N. the limiting melting-point is about 450° C. The primary explosives such as lead azide, lead styphnate, mercury fulminate and tetrazene all show the same effect except that the limiting temperature is somewhat higher. Some results for mercury fulminate and tetrazene are given in Table XV.

TABLE XV. *Initiation of mercury fulminate and tetrazene by impact in the presence of grit*

| Grit added | Hardness (Mohs' scale) | Melting-point (° C.) | Explosion efficiency (%) | |
|---|---|---|---|---|
| | | | Mercury fulminate 240 g. 35 cm. | Tetrazene 240 g. 10 cm. |
| Nil | — | — | 0 | 0 |
| Silver nitrate | 2–3 | 212 | 0 | 0 |
| Potassium nitrate | 2–3 | 334 | — | 3 |
| Potassium dichromate | 2–3 | 398 | — | 0 |
| Silver bromide | 2–3 | 434 | 0 | 31 |
| Lead chloride | 2–3 | 501 | 0 | 30 |
| Silver iodide | 2–3 | 550 | 70 | 80 |
| Borax | 3–4 | 560 | 100 | 100 |
| Bismuthinite | 2–2·5 | 685 | 100 | 100 |
| Chalcocite | 3–3·5 | 1100 | 100 | 38 |
| Galena | 2·5–2·7 | 1114 | 100 | 100 |
| Calcite | 3 | 1339 | 100 | 38 |

As mentioned earlier the size of the grit particles is of some importance and particles of about $20\mu$ in diameter are more effective than smaller ones. Nevertheless there is evidence that very small particles can have a sensitizing effect. *Williams* (1950) has observed that particles of diamond dust as small as $0.5\,\mu$ could still have a sensitizing effect on a number of solid explosives. Particles as small as this are present in most explosives, however carefully they are prepared.

### 4.4. Size of hot spot

Some estimates have already been given of the size of hot spots necessary to initiate explosion. *Bowden, Stone* and *Tudor* obtained a value of $10^{-2}$ to $10^{-3}$ cm. for the diameter of hot spots in the initiation by friction of explosion in nitroglycerine. Rideal and Robertson (1948 *a*) have made some theoretical calculations on the critical size of the hot spot required at different initial temperatures. Their assumption is that explosion spreads from the hot spot by self-heating, namely, when the heat evolved during the decomposition is greater than the heat lost by conduction. Under such conditions, the hot spot grows and the explosion is not extinguished. The hot spot is assumed to be spherical and to consist of liquid explosive. The rate at which the reaction occurs in the hot spot is calculated from the Arrhenius relation for the thermal decomposition and the heat of reaction is assumed to be the same as for the thermal decomposition. The heat lost is calculated using Fourier's relation for heat conduction assuming the thermal conductivity and diffusivity of the liquid and solid explosive to be the same.

Consider a spherical hot spot to be formed at zero time. The temperature at a point outside the hot spot distant $r$ from the centre will have increased after a time $t$ by $\theta$. The Fourier equation is

$$\frac{\partial \theta}{\partial t} = \frac{k}{\rho c}\left(\frac{\partial^2 \theta}{\partial r^2} + \frac{2}{r}\frac{\partial \theta}{\partial r}\right), \tag{4.1}$$

where $k$, $\rho$ and $c$ are the thermal conductivity, density and specific heat of the explosive respectively. If the radius of the hot spot itself is $a$ and at zero time it is at a uniform temperature $\theta_0$ above its surroundings, the boundary conditions for (4.1) are

$$\theta = 0, \quad \text{when} \quad t = 0 \quad \text{for} \quad r > a,$$
$$\theta = \theta_0, \quad \text{when} \quad t = 0 \quad \text{for} \quad 0 < r < a,$$

and these enable the equation to be solved and an explicit expression for the temperature at any point to be calculated.

After a small interval of time $t$, a concentric spherical shell of thickness $dr$ at a distance $r$ ( $> a$) from the centre of the hot spot has gained an amount of heat $4\pi r^2 \theta \rho c\, dr$. Hence the heat lost from the hot spot equals

$$\int_a^\infty 4\pi r^2 \theta \rho c\, dr. \tag{4.2}$$

In finding explosion conditions the heat produced by reaction must be introduced:

$$\text{Heat evolved} = \tfrac{4}{3}\pi a^3 \rho tq\, Ae^{-E/RT}, \tag{4.3}$$

where $q$ is the heat of reaction per unit mass and $Ae^{-E/RT}$ the Arrhenius relation for the rate of reaction. From these equations a critical value of the hot-spot temperature $T^\circ$ K. may be calculated for a given size of hot spot where the heat evolved just balances the heat lost.

Some values for different explosives are given in Table XVI.

TABLE XVI. *Critical hot-spot temperatures,* $^\circ$ C.

| Explosive | Hot-spot radius | | |
|---|---|---|---|
| | $10^{-3}$ cm. | $10^{-4}$ cm. | $10^{-5}$ cm. |
| P.E.T.N. | 350 | 440 | 560 |
| Cyclonite | 385 | 485 | 620 |
| Tetryl | 425 | 570 | 813 |

These calculations give a value of $c$. $10^{-3}$ to $10^{-4}$ cm. for the size of the hot spot necessary for explosion, in good agreement with the value obtained by *Bowden et al.* (1947).

## 4.5. Initiation by impact of explosion in primary explosives

We see that the general behaviour of the primary explosives show some similarities to that of the secondary explosives. Grit has a sensitizing effect and the limiting hot-spot temperature is in the region of 500° C. (Table XV, § 4.3). It is also found that the time from the first moment of impact to the onset of explosion with substances such as lead azide, mercury fulminate and lead styphnate is, in the presence of grit, about 50–100 $\mu$sec. This is of the same order as that observed with P.E.T.N. The main difference is that with the primary explosives friction between the crystal particles themselves can initiate the explosion. For this reason the compression of gas spaces can be less important, and experiments with tetrazene and cyanuric triazide support this view (*Bowden* and *Williams*, 1951 *a*).

## 4.6. Hot spots in the ignition of gunpowder

Even with a heterogeneous material such as gunpowder there is evidence that the mechanism of initiation by impact is again due to hot-spot formation. As mentioned earlier *Blackwood* and *Bowden* (1951) have shown that the ignition of gunpowder is due to the compression of included gas. They also find that, if the ignition is sensitized by grit particles, the controlling factor is the melting-point of the particle (see Table XVII).

TABLE XVII. *The ignition of gunpowder by impact in the presence of grit*

Mass of striker 1800 g.   Height of fall 50 cm.

| Grit added | Melting-point (° C.) | Initiation (%) |
|---|---|---|
| Ammonium sulphate | 100 | 0 |
| Zinc octacosanoate | 130 | 20 |
| Ammonium thiocyanate | 150 | 30 |
| Ammonium nitrate | 170 | 60 |
| Silver nitrate | 212 | 100 |
| Sodium nitrate | 307 | 80 |
| Sodium acetate | 324 | 80 |
| Potassium nitrate | 334 | 100 |
| Sodium bromate | 381 | 100 |
| Cadmium iodide | 388 | 100 |
| Silver chloride | 455 | 80 |

It is evident that the limiting temperature with gunpowder ( < 130° C.) is very much lower than that observed with explosive materials. This temperature is close to the melting-point of sulphur.

## CHAPTER V

# THE GROWTH OF EXPLOSION TO DETONATION

In this chapter we shall discuss the way in which the explosion grows from the initiating hot spot and develops into a large-scale detonation. The experiments were carried out mainly with thin films of explosive about 0·1–0·5 mm. thick, and the development of the explosion in the immediate vicinity of the point of initiation was followed by specially designed electronic and photographic apparatus. Initiation was effected either by the compression of a gas bubble, by friction on a particle of grit, by spark or by a hot wire. It is shown that with all the secondary explosives the reaction begins as a comparatively slow burning which accelerates, and after travelling a short distance passes over into a low velocity detonation. Some primary explosives such as mercury fulminate and lead styphnate behave in the same way. In general, however, the burning stage is absent (or is too short to be observed) for the azides and fulminates, and detonation sets in very close to the point of initiation.

## 5.1. Types of camera used

A convenient method of following the growth of explosion is by a rotating drum camera. A simple type is illustrated in figs. 25 and 29 in Chapter III. Many investigators have developed such instruments in order to obtain accurate values for detonation velocities in large charges of explosive; see, for example, Lafitte (1925) and Jones (1928). Other cameras of the mirror type have been used by Payman et al. (1937), Cairns (1944), Herzberg and Walker (1948), etc. One of the cameras used in our investigations is shown in fig. 38. The film is mounted on the inside of the drum and a prism fixed inside the camera throws the image of the explosion on the film. The camera is focused on the slit, set so that it gives an image on the film at right angles to the direction of motion. The explosive is spread on a transparent material (glass or mica) placed over the slit. When an explosion takes place the flame spreads from the point of initiation, and from the angle the flame makes to the horizontal on the film it is possible to calculate

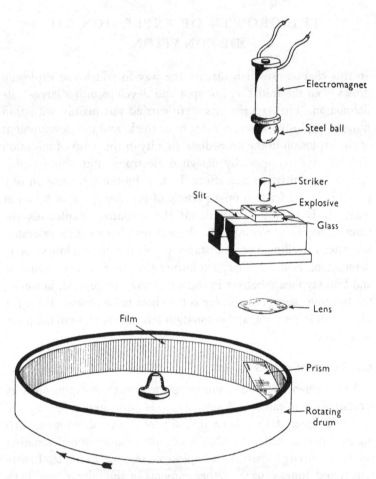

Fig. 38. Diagrammatic representation of high-speed drum camera and explosion apparatus.

the velocity of propagation. The speed of the film with this camera is $c$. 85 m.sec.$^{-1}$, and it has a resolution down to a microsecond. It is sometimes necessary to use much higher speeds, and here a camera involving a new principle which has been developed by *Courtney-Pratt* (1949) may be used. This camera makes use of

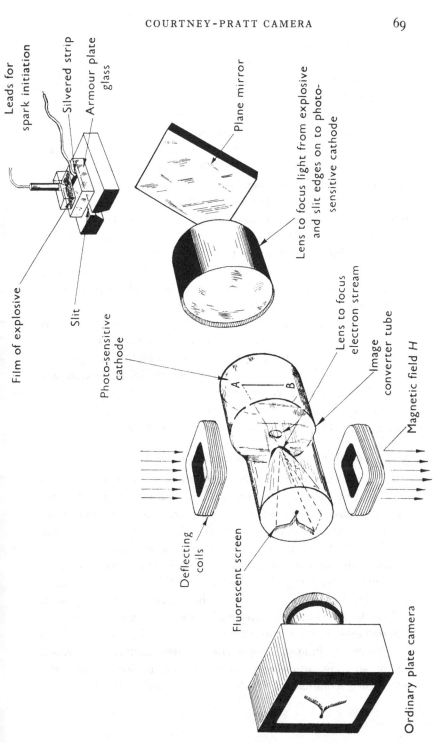

Fig. 39. Schematic diagram of the image converter tube as used for photographing an explosion.

an image converter tube and suitable electronic apparatus for deflecting the electron stream. The advantages of this camera over the conventional moving film or mirror type are considerable. Extremely high writing speeds ( > 2000 m.sec.$^{-1}$) may be obtained and the instrument has a high aperture and high resolution. A diagrammatic sketch of the apparatus is given in fig. 39. If the photosensitive surface of the image converter tube is illuminated along a line $AB$, a bright line will be formed on the fluorescent screen. If an electric or magnetic field is introduced to act on the electron stream between the focusing electron lens system and the fluorescent screen then the position of the line on the screen can be altered at will. If instead of uniform illumination along $AB$, a bright point image is made to traverse linearly along $AB$, and further a vertical magnetic field is applied to the tube which also changes linearly with time, an inclined line is obtained and the inclination of this line gives a measure of the relative velocities of the image along $AB$ and the rate of sweep of the time base. It is possible to measure the velocity of any luminous phenomenon moving along $AB$ with this apparatus, and resolutions of the order of $10^{-9}$ sec. can be obtained. The camera can also be used to take a rapid series of still photographs.

*Courtney-Pratt* has recently developed a different technique for taking a rapid series of stills; see also Sultanoff, 1950.

## 5.2. Growth of explosion in liquids

(1) *Explosions obtained with cavity strikers.* If an explosion is produced by allowing a cavity striker to fall into a lens of nitroglycerine placed on a metal anvil, it is found that a pattern or blast mark is imprinted on the metal anvil. Fig. 40*e* shows the type of blast mark that is produced on a brass anvil which had been coated with a thin film of lead. The circular imprint corresponds to the rim of the cavity. The lead film within the cavity is little damaged, but the eruption under the cavity rim is clearly seen. When the metal anvil is replaced by a sheet of mica placed on a photographic plate, a stationary picture of the light effects within the cavity is obtained. Some characteristic pictures are shown in fig. 40. Fig. 40*a* shows clearly the uneven illumination and incomplete combustion which may occur within the cavity. In

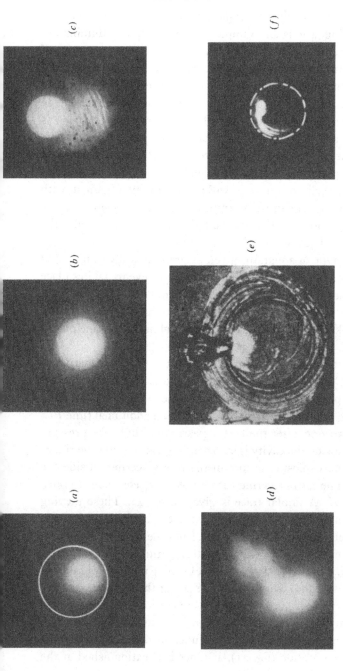

(a) Impact on photographic plate covered with mica. Note non-uniform illumination in cavity (×18).
(b) As for (a) showing uniform illumination of the cavity and absence of any exterior luminous effects (×13).
(c) Strong eruption from cavity (×11).
(d) Nitroglycol light pattern. Observe pronounced eruption from cavity (×10).
(e) Brass anvil (lead plated) after a high energy cavity impact. Note pronounced eruption channel breaking through the wall of the cavity (×23).
(f) Blasting gelatine. Light pattern showing uneven distribution of luminous effects within cavity, outline of cavity indicated by dotted line (×16). Cf. (a).

Fig. 40. Explosion initiated by cavity impact. Luminous and blast effects obtained with nitroglycerine.

this case it is obvious that the explosion has not propagated beyond the cavity. Fig. 40*b* is an example of more complete illumination within the cavity itself. Fig. 40*c* and *d* show typical examples of eruption from the cavity. The flame blasts through the wall of the cavity into the surrounding explosive, and since the surrounding explosive in this case is not confined, it is blown away and the propagation does not proceed far. If the surrounding explosive is confined as a thin film, the eruption causes propagation of the explosion throughout the film. The resemblance between the blast marks and the light photographs is obvious. Fig. 40*f* is a good example of localized burning within the cavity obtained with blasting gelatine. Again the burning is confined to a small region of the cavity, and, in this case, the light photograph provides the only evidence that initiation has occurred.

Using the type of apparatus given diagrammatically in fig. 25, it has been possible to examine the sequence of events taking place within the cavity itself, although it has a radius of only *c*. 0·4 mm. The slit system is at right angles to the axis of rotation of the drum. A cavity striker is allowed to fall into a lens of nitroglycerine placed over the slit.

Fig. 41 is a magnified picture. The light within the cavity first begins at *A* and spreads slowly inside the cavity for a distance which is approximately equal to the diameter of the cavity *BB*. The burning within the cavity proceeds for an additional time *BC*. At *C* some 50 μsec. after the first appearance of light the pressure of the gases inside the cavity is sufficiently great to burst the rim of the cavity and explosion occurs in the nitroglycerine outside the cavity *CD*. The nitroglycerine within the cavity continues to burn for a time *CE*. A similar trace is given in fig. 42. These moving film pictures should be compared with the stationary pictures, fig. 40, which also show the eruption from the cavity.

A large number of similar experiments show this sequence of events to be general. Frequently, however, the burning merely flares up and dies down inside the cavity without breaking through (see fig. 43). The explosion fails to grow and if it were not for the photographic evidence there would be little or no sign that initiation has occurred.

Though the burning usually continues in the cavity for some time after the eruption (fig. 41), at times it is extinguished at the

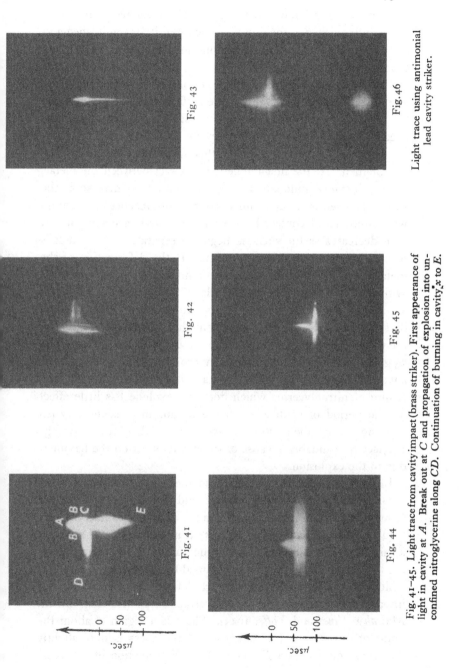

Fig. 43

Fig. 46

Light trace using antimonial lead cavity striker.

Fig. 42

Fig. 45

Fig. 41

Fig. 44

Fig. 41–45. Light trace from cavity impact (brass striker). First appearance of light in cavity at *A*. Break out at *C* and propagation of explosion into unconfined nitroglycerine along *CD*. Continuation of burning in cavity, *x* to *E*.

moment the eruption occurs (see fig. 44). Occasionally, instead of being extinguished at the eruption, the intensity is diminished and it flares up again before finally going out. Fig. 46 is an extreme example of this.

The photographic observations described above have been amplified by an electrical measurement of the conductance between the impacting metal surfaces. It is possible to study by means of a cathode-ray oscillograph the rapidly changing electrical conductance during impact, and from this the duration of impact may be obtained. At the first moment of contact between the metals the conductance suddenly rises. It continues to increase as the surfaces come into more intimate contact and reaches a maximum when the area of contact between the metals is a maximum. It then decreases as the surfaces begin to separate and again falls to zero when separation is complete. If the deformation of the metals were completely elastic the conductance curve would be symmetrical, but if there is some plastic deformation it will show an asymmetry. The general shape of the curves obtained is illustrated in fig. 47, and it is evident that plastic deformation of the metal does occur. When a curved brass cavity striker weighing 35 g. is allowed to fall 10 cm. (energy 350 g.cm.) on to a clean brass anvil, the average period of contact is found to be 160 $\mu$sec. A film of nitroglycerine which does not explode has little effect on the period of contact or on the minimum resistance. When explosion occurs the period of contact is somewhat lower—average 130 $\mu$sec. presumably because of the lifting effect on the hammer, due to the explosion.

In fig. 47 a sketch has also been included of the luminous effects obtained on the moving film camera, and the relation between the sequence of the luminous effects and the period of contact is clear.

(2) *Explosions between flat surfaces.* If nitroglycerine is spread as a thin film on a flat anvil and is struck with a flat brass hammer, the type of blast pattern that is obtained may be seen in Plate I(a) which shows a typical picture of marks left on a brass hammer after impact on a number of nitroglycerine drops (*Bowden, Eirich, Mulcahy, Vines* and *Yoffe*, 1943). There is a region A about the point of initiation J (see later) where the brass is only slightly damaged and a region B where severe deformation of the brass

PLATE I

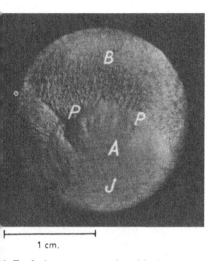

(a) Explosion pattern produced by impact on number of nitroglycerine drops. Striker of brass.

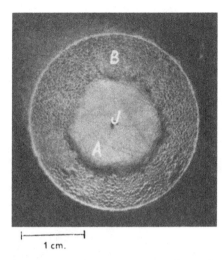

(b) Pattern produced by explosion in a thin film of nitroglycerine. Initiation by spark. Anvil of brass.

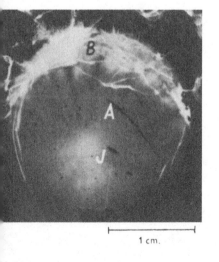

(c) Light pattern on a photographic plate. Initiation by impact on a thin film of nitroglycerine containing air bubbles.

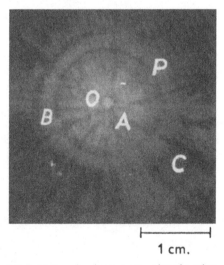

(d) Initiation by impact on nitroglycerine drops. Light photograph of explosion on a plastic support. Radial lines are due to the cracking of the support.

has occurred. The transition from slight deformation to heavy deformation is sudden. Just inside the region *B* there is a band of unmarked brass *PP* where even the polished Beilby layer has not been removed. A similar type of pattern is obtained if the explosion is initiated by the passage of a condenser spark discharged through a thin film of nitroglycerine confined between two parallel flat brass surfaces (see Plate I(*b*)). The point of initiation is at *J* where the spark passed, and the region of untouched brass *A* suddenly transforms to a region of 'heavy blasting' *B*.

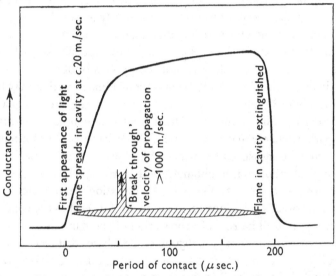

Fig. 47. Diagrammatic sketch of the period of contact of a brass cavity striker on a brass anvil. Incidence of luminous effects have also been included (cf. fig. 41).

When explosion occurs on a mica sheet placed on a photographic plate or on a transparent surface at the focus of a camera lens, the light effects may also be recorded. Plate I(*c*) is a light pattern obtained from the explosion of a thin film of nitroglycerine containing air bubbles and subjected to impact. The luminous effects may be divided into two regions *A* and *B* about the point of initiation *J*. The patterns of the light effects correspond very closely to the blast patterns obtained on metal surfaces even to the dark band *PP* corresponding to the undamaged metal in Plate I(*a*). In Plate I(*d*) a light pattern obtained by *Mulcahy* and *Vines* (1947) is given of an explosion initiated by spark and taken by a camera focused on the transparent anvil.

It is clear that when explosion is initiated in a thin film of nitro-glycerine between two flat surfaces by spark or by impact in the presence of a gas space it develops in two stages. The first stage is a deflagration which spreads out radially for a distance of some 5 mm. and is then suddenly transformed into a more violent decomposition. This second stage emits more light and it also deforms brass by forming numerous small pits in the metal.

In another series of experiments the nitroglycerine is spread as a ring, with the centre of the ring immediately above the slit, and the explosive is then hit with a flat hammer. In this way the small air bubble trapped in the explosive lies above the slit, so that the source of initiation, which is the hot compressed bubble, is located directly above the slit. For thin continuous films initiation may be brought about by placing a particle of grit over the slit and again subjecting the film to impact. Other methods such as a hot wire or a spark discharge may also be used. The velocity of the luminous effects may be measured by means of a rotating drum camera.

A typical photograph obtained with nitroglycerine is shown in fig. 48. $A$ represents the point of initiation from which the explosion spreads at an increasing speed until it reaches the edge of the striker at $B$. The rate of propagation of the flame in this case accelerates from about 180–650 m.sec.$^{-1}$. The distance $BB$ is about 2·5 cm. In many cases the accelerating burning passes over into detonation, e.g. at $B$ in fig. 49, and continues with a constant velocity of about 2200 m.sec.$^{-1}$.

Another method of initiating an explosion by impact is to use a striker containing a small cavity. The nitroglycerine is spread as a film over the slit and a flat striker with a hemispherical cavity in the centre is allowed to fall on the film, the cavity being immediately over the slit during impact. The results are shown in figs. 50 and 51. The light of explosion first appears in the cavity at $A$ and the burning continues in the cavity for the time $AB$. At $B$ the explosion bursts out of the cavity and one of two things may happen. The explosion may either develop from the eruption as a rapid accelerating combustion up to 400 m.sec.$^{-1}$ which suddenly passes over to the 2000 m.sec.$^{-1}$ detonation (fig. 50), or detonation may

PLATE II

(a) Low velocity detonation in thin film of nitro-glycerine initiated by spark. Point of initiation not visible.

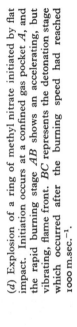

(c) As for (a).

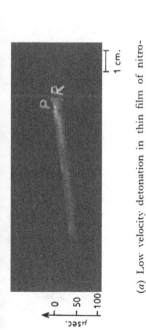

(b) 'Across the slit' photograph obtained with drum camera.

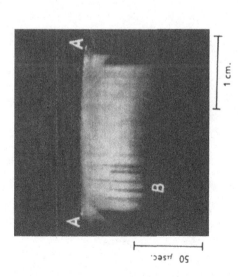

(d) Explosion of a ring of methyl nitrate initiated by flat impact. Initiation occurs at a confined gas pocket A, and the rapid burning stage AB shows an accelerating, but vibrating, flame front. BC represents the detonation stage which occurred after the burning speed had reached 1000 m.sec.$^{-1}$.

set in quite close to the eruption (fig. 51). This behaviour is similar to that observed by *Bowden* and *Gurton* for spark-initiated explosion (figs. 52 and 53), initiation by a few grains of azide fired by a spark (fig. 54), or when a hemispherical cavity striker is allowed to fall on to a film of nitroglycerine confined between two flat surfaces, the upper surface having a narrow hole drilled through it to allow the cavity striker to fall through it (fig. 55).

There is some evidence for a dark space between the two stages (*Mulcahy* and *Vines* (1947)) suggesting that detonation may set in ahead of the combustion, but further work on this is necessary.

*The low-velocity detonation in nitroglycerine.* It seems probable that the velocity of propagation of the second-stage process, 2000 m.sec.$^{-1}$, corresponds to that of the so-called low-velocity detonation obtained when a tube of nitroglycerine is initiated by a weak detonator. Chariton and Ratner (1943) find that when nitroglycerine is confined in a glass tube of less than 2 mm. bore, the low-velocity detonation (*c.* 1800 m.sec.$^{-1}$) is obtained. In tubes of diameter greater than about 5 mm. the detonation may proceed at either of two velocities, viz. about 1800 m.sec.$^{-1}$ or about 7500 m.sec.$^{-1}$.

*Mulcahy* and *Vines* have made a photographic study of the 2000 m.sec.$^{-1}$ detonation stage in thin layers of nitroglycerine. Two observations may be made from the photographs, Plate II *a* and *c*. The first is connected with the duration of the luminosity and motion of the gaseous products left in the wake of the detonation. In Plate II *a* it is evident that products of decomposition remain luminous for an appreciable time after the passage of the detonation through the explosive. Furthermore, after a perceptible motion forward the hot gases move backwards from the detonation zone with a velocity of about 400–800 m.sec.$^{-1}$. This motion of the gaseous products resembles those obtained by many workers who investigated the propagation of detonation in combustible gas mixtures (see Jost, 1946). The second observation is that the detonation process may be discontinuous. The pit markings characteristic of the detonation process, Plate I *a* and *b*, suggest that this stage is made up of a succession of localized explosions about 0·5 mm. apart. The camera traces sometimes show variations in luminous intensity along their length which may be ascribed to

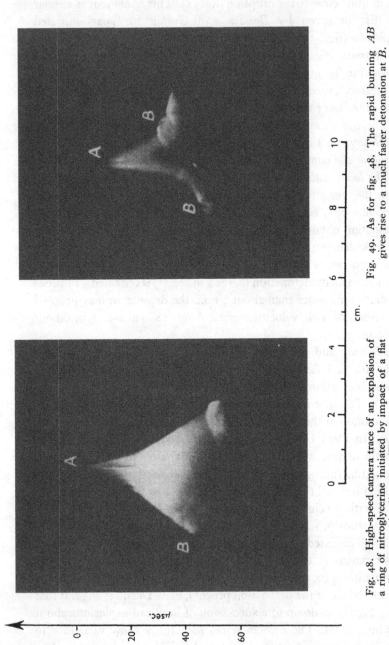

cm.

Fig. 48. High-speed camera trace of an explosion of a ring of nitroglycerine initiated by impact of a flat striker 2·5 cm. in diameter. The point of initiation $A$ corresponds to a compressed air bubble and the explosion $AB$ spreads as a rapid accelerating burning at 180–650 m.sec.$^{-1}$.

Fig. 49. As for fig. 48. The rapid burning $AB$ gives rise to a much faster detonation at $B$.

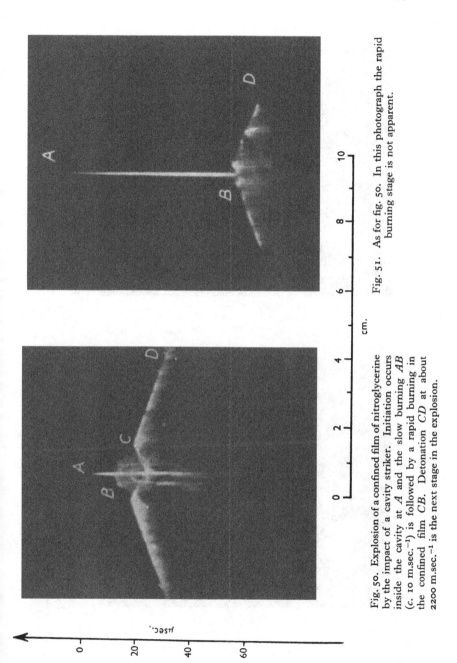

Fig. 50. Explosion of a confined film of nitroglycerine by the impact of a cavity striker. Initiation occurs inside the cavity at $A$ and the slow burning $AB$ (c. 10 m.sec.$^{-1}$) is followed by a rapid burning in the confined film $CB$. Detonation $CD$ at about 2200 m.sec.$^{-1}$ is the next stage in the explosion.

Fig. 51. As for fig. 50. In this photograph the rapid burning stage is not apparent.

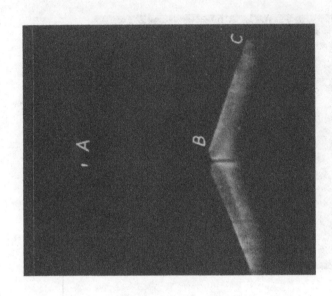

Fig. 53. As for fig. 52. There is no initial burning region visible in this experiment.

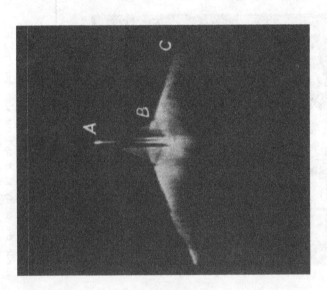

Fig. 52. Explosion of confined nitroglycerine initiated by an electric spark *A*, showing the rapid burning stage *AB* followed by detonation *BC*.

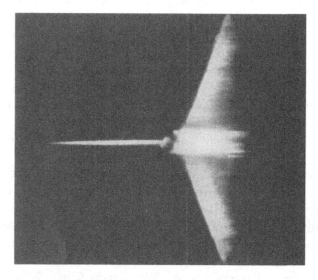

Fig. 55. Photograph similar to fig. 50 using a hemispherical ended cavity striker.

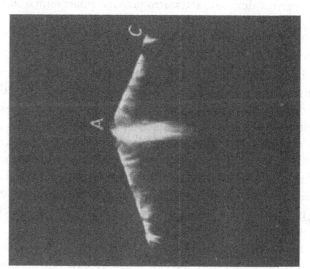

Fig. 54. Explosion of a confined film of nitroglycerine initiated by the detonation of a few crystals of lead azide at *A*. Only the detonation stage *AC* is observed.

this phenomenon. When the explosion propagates diametrically across the slit instead of along it, the result obtained is shown in Plate II b. The detonation front first becomes visible to the camera at $A$ and passes across the slit out of view, but the products remain luminous for the time $AB$. The line of light $AA$ is a trace of the detonation front. This line is finely serrated and is made up of a number of bright spots separated by spaces of lower luminosity. These bright spots may be identified with the pits produced on the metal surface during detonation. When stable detonation is not set up the camera traces show that the explosion proceeds in a more markedly discontinuous fashion; see Plate II c.

(3) *Growth of explosion in liquids other than nitroglycerine.* Other liquid explosives such as methyl nitrate behave in a similar manner to nitroglycerine. Plate II d shows the growth of explosion in methyl nitrate when spread as a ring and hit by a flat hammer. There is again the rapid and accelerating burning from the point of initiation which transforms after about $20\,\mu$sec. to a detonation at $B$ (*Bowden* and *Gurton*, 1949). *Vines* (1947) found the same behaviour in the spark initiation of nitroglycol, diglycol tetranitrate, diethylene glycol dinitrate, tetranitromethane-toluene mixtures.

## 5.3. Growth of explosion in solid secondary explosives

(1) *Propagation in P.E.T.N.*

A solid explosive such as P.E.T.N. may be exploded in the same way as nitroglycerine. The point of initiation may be located either by the presence of a small particle of grit in the film of explosive, or by spreading the P.E.T.N. as a ring so that the trapped gas space is over the slit. The explosion trace obtained is shown in fig. 56. In this experiment initiation occurs at $A$ where a small speck of glass had been placed in the centre of the film of P.E.T.N. From this point the flame spreads in both directions at fairly low speeds (100–400 m.sec.$^{-1}$) and usually accelerates as it approaches the edges of the impacted area.

The second method of initiation is to spread the P.E.T.N. as a ring, with the centre over the slit. Figs. 57 and 58 show traces obtained with such a layer of P.E.T.N. about 0·1 mm. thick. Initiation begins at the gas space at $A$ and the flame spreads from this point at a comparatively slow rate. In fig. 58 the initial rate of burning $AB$ accelerates from 300 to 1000 m.sec.$^{-1}$. It might be expected that this rate would be dependent on the experimental conditions and the rate varies from 100 to 1000 m.sec.$^{-1}$ After a short distance at the point $B$ this burning changes over into a detonation, the speed of which varies from 1000 to 2500 m.sec.$^{-1}$ according to the density and film thickness of the explosive. In fig. 58 the second stage $BC$ is spreading at 1620 m.sec.$^{-1}$. The transition from low-speed to high-speed propagation usually occurs at obvious physical discontinuities, that is, at the edge of the hammer, but this is not always the case. The traces are very similar to those observed with nitroglycerine, and in some cases a dark space may be seen at the transition from the burning to detonation.

It is possible to initiate an explosion in a film of P.E.T.N. by a spark provided the explosive is heavily constrained. The trace given in fig. 59 is similar to those obtained in impact experiments.

(2) *Propagation in cyclonite and tetryl.*

Films of cyclonite and tetryl give similar results but the initial burning rate is slower (100–300 m.sec.$^{-1}$) and the explosion does not propagate to the thin film of explosive surrounding the hammer. The propagation rate is slow and the flame persists for a considerable time.

In attempting to propagate these explosions beyond the limits of the striker radius, results similar to those shown in figs. 60 and 61 are obtained. In neither case is the transformation from deflagration to detonation possible. There is some evidence of this continuation of the deflagration in the case of cyclonite, and in fig. 60 where $AB$ represents the slow burning under the hammer, $BC$ represents a continuation of the burning into the confined area. This flame is soon extinguished and the majority of the cyclonite outside the striker is unconsumed by the explosion.

It is, however, possible to detonate films of cyclonite or tetryl if a small amount of lead azide containing a particle of grit is spread under the striker and used as the source of initiation (see fig. 62).

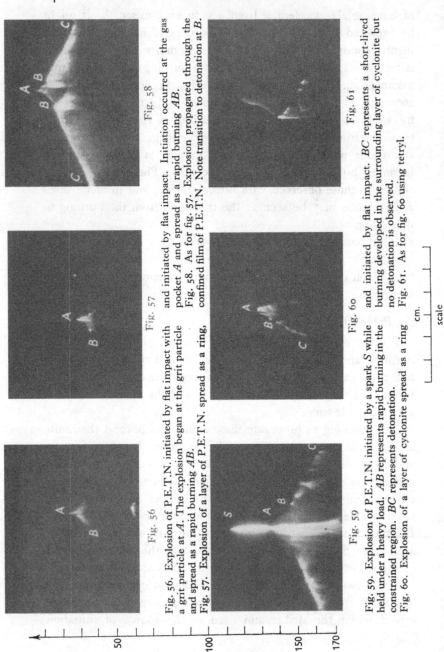

Fig. 56

Fig. 57

Fig. 58

Fig. 59

Fig. 60

Fig. 61

Fig. 56. Explosion of P.E.T.N. initiated by flat impact with a grit particle at A. The explosion began at the grit particle and spread as a rapid burning AB.

Fig. 57. Explosion of a layer of P.E.T.N. spread as a ring, and initiated by flat impact. Initiation occurred at the gas pocket A and spread as a rapid burning AB.

Fig. 58. As for fig. 57. Explosion propagated through the confined film of P.E.T.N. Note transition to detonation at B.

Fig. 59. Explosion of P.E.T.N. initiated by a spark S while held under a heavy load. AB represents rapid burning in the constrained region. BC represents detonation.

Fig. 60. Explosion of a layer of cyclonite spread as a ring and initiated by flat impact. BC represents a short-lived burning developed in the surrounding layer of cyclonite but no detonation is observed.

Fig. 61. As for fig. 60 using tetryl.

cm.

scale

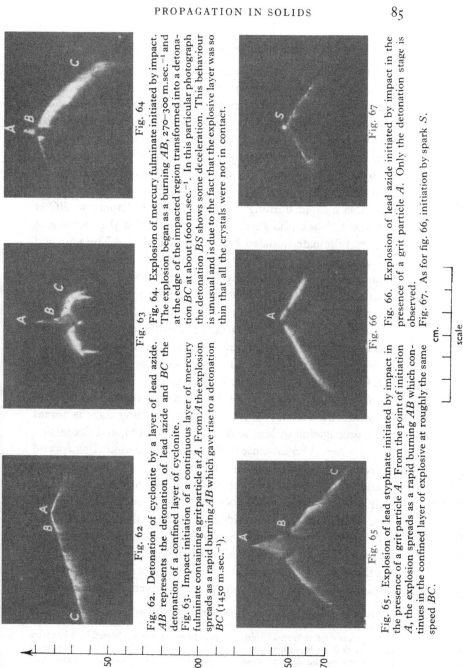

Fig. 62

Fig. 63

Fig. 64

Fig. 65

Fig. 66

Fig. 67

Fig. 62. Detonation of cyclonite by a layer of lead azide. $AB$ represents the detonation of lead azide and $BC$ the detonation of a confined layer of cyclonite.

Fig. 63. Impact initiation of a continuous layer of mercury fulminate containing a grit particle at $A$. From $A$ the explosion spreads as a rapid burning $AB$ which gave rise to a detonation $BC$ (1450 m.sec.$^{-1}$).

Fig. 64. Explosion of mercury fulminate initiated by impact. The explosion began as a burning $AB$, 270–300 m.sec.$^{-1}$ and at the edge of the impacted region transformed into a detonation $BC$ at about 1600 m.sec.$^{-1}$. In this particular photograph the detonation $BS$ shows some deceleration. This behaviour is unusual and is due to the fact that the explosive layer was so thin that all the crystals were not in contact.

Fig. 65. Explosion of lead styphnate initiated by impact in the presence of a grit particle $A$. From the point of initiation $A$, the explosion spreads as a rapid burning $AB$ which continues in the confined layer of explosive at roughly the same speed $BC$.

Fig. 66. Explosion of lead azide initiated by impact in the presence of a grit particle $A$. Only the detonation stage is observed.

Fig. 67. As for fig. 66, initiation by spark $S$.

## 5.4. Growth of explosion in primary explosives

Primary explosives fall into two classes in their mode of propagation. Lead styphnate, sodium fulminate and mercury fulminate resemble the secondary explosives, and the explosion develops from the point of initiation as an accelerating burning which may pass over into a detonation. Essentially similar results are obtained when initiation occurs by impact, by spark, or by a hot wire.* Fig. 63 is a picture obtained with mercury fulminate initiated by impact when a grit particle is present over the slit. The explosion spreads as a rapid burning $AB$ followed by a detonation $BC$ at 1450 m.sec.$^{-1}$ (see also Patry (1933) who studied the burning of mercury fulminate initiated by hot wires and also Muraour and Wohlgemuth (1936) who find that the burning stage with mercury fulminate disappears if the fulminate is heated to $c.$ 100° C. before initiation). Fig. 64 shows another picture obtained with mercury fulminate. The burning stages observed with lead styphnate is shown in fig. 65. The traces obtained with lead styphnate are unusual in that the burning continues into the unimpacted area without any sharp change in velocity and at an irregular speed. It must be assumed, therefore, that either deflagration continues without detonation being set up or that the detonation velocity is very low.

*Bowden* and *Gurton* (1949) find that the burning stage is absent in the propagation of lead azide (figs. 66 and 67), and detonation at 1500 m.sec.$^{-1}$ begins close to the point of initiation. *Bowden* and *Williams* (1951) have examined the behaviour of a large number of azides and fulminates in an endeavour to correlate the impact sensitivity and detonation velocity with the molecular structure of these compounds. Lithium azide, calcium azide, thallous azide, silver azide, thallous fulminate, silver fulminate, cadmium fulminate and cuprous fulminate, all behave similarly to lead azide when confined as a thin film.† Values for the detonation velocities of some azides spread as a thin film $c.$ 0·1 mm. thick are given in Table XVIII.

---

* An account of the ignition of deflagrating solids by means of hot wires is given by Jones (1949) and Stout (1950).

† Cyanuric triazide and trinitro triazido benzene when initiated by a hot wire in an unconfined film burn without detonation with a velocity of several metres per second in the case of cyanuric triazide and of several centimetres per second for trinitro triazido benzene.

For the alkali and alkaline earth azides the influence of the cation radius is clear. Sodium azide does not detonate under impact or even when some lead azide is exploded near it. Lithium azide, on the other hand, does explode. The same relation exists between barium and calcium azides. This behaviour may be associated with the greater covalent character of the lithium and calcium azides.

TABLE XVIII

| Azide | Velocity of detonation in m.sec.$^{-1}$ | Cation radius |
|-------|-------------------------------------|---------------|
| NaN$_3$ | — | 0·95 |
| LiN$_3$ | 990 | 0·60 |
| Ba(N$_3$)$_2$ | — | 1·35 |
| Ca(N$_3$)$_2$ | 770 | 0·99 |
| TlN$_3$ | 1482 | 1·44 |
| AgN$_3$ | 1695 | 1·13 |
| Pb(N$_3$)$_2$ | 2320 | 1·21 |

The absence of a burning region in the case of the inorganic azides may be associated with the fact that the decomposition takes place in a small number of steps to give a metal and nitrogen. The heat is thus liberated sufficiently rapidly to give rise to a detonation. When the possibility of step-wise decomposition exists the heat liberated in each step may be sufficient to support only a burning region. Such is in fact the case with complex organic azides. For example, trinitro triazido benzene and cyanuric triazide burn when they are ignited as an unconfined film (*Bowden* and *Williams*, 1951 a). The trinitro triazido benzene around the hot spot decomposes first into hexanitroso benzene and nitrogen and the hexanitroso benzene decomposes at a later stage. Explosion only occurs with these organic azides when they are confined or when they are ignited in a definite pressure of an inert gas such as nitrogen. *Yoffe* (1951) has suggested that the conditions leading to explosion depend on the self-heating brought about by the recombination of atomic or active nitrogen near the surface of the azide, and that the effect of confinement or of an inert gas is to prevent the escape of nitrogen atoms or active nitrogen molecules away from the surface.

## 5.5. Nature of the initial burning region

Most investigators have found that the burning rate of propellants and some explosives is proportional to the pressure. During impact very high pressures are developed, but they are unlikely to exceed the flow pressure of the metal. If Andreev's (1946) figures for the burning speed of nitroglycerine are extrapolated to $10^4$ atm., which corresponds to the flow pressure of brass, a theoretical burning speed of 10 m.sec.$^{-1}$ is obtained which is of the same order as the speed observed inside the brass cavity strikers. However, burning speeds up to 400–600 m.sec.$^{-1}$ have been observed for nitroglycerine. This suggests that the whole front is undergoing a mass movement which is superimposed on the true burning speed. In other words the products and the explosive are being forced away from the centre of explosion.

In this mass movement as the explosive accelerates it may attain speeds approaching the velocity of sound in the thin film, and will therefore build up a shock front. It also attains a speed of the same order and same direction as the streaming velocity of the products behind a low velocity detonation wave. The development of low velocity detonation from the rapid movement of the flame front is therefore not surprising. A somewhat similar mechanism for the development of detonation in a large mass of explosive has been suggested by Kistiakowsky (see below). A similar argument has been applied to solid explosives by *Bowden* and *Gurton* (1949) and by Ubbelohde (1950) (see also Muraour, 1950).

'*Critical size*' *and detonation of explosives.* In a well-confined explosive film, the transition from burning to detonation occurs within a few millimetres of the point of initiation. Kistiakowsky (1948) has described experiments on the deflagration of explosives and concludes that every deflagrating explosive can be made to detonate under suitable conditions. He postulates the formation of shock waves in the flowing gases produced during burning, and that in a mass of explosive above a 'critical size' the shock will be strong enough to change the burning over to a detonation.

Kistiakowsky considers the case of a mass of solid granular explosive ignited at a localized region within the bulk. As the explosive burns the gases formed cannot readily escape between the spaces left between the explosive crystals and pressure gradients are produced. The gas pressure therefore increases, resulting in an increase in the burning velocity which in turn produces a further increase in the rate of pressure rise. The gases therefore stream away from the centre of deflagration at a constantly rising backing pressure and temperature, i.e. with increasing velocity. These are conditions which favour the formation of shock waves. The shocks start the rapid deflagration of the explosive layers through which they pass and the energy so released reinforces the shock. The shocks finally reach an intensity where the entire explosive is consumed in their passage and the entire available energy can be used for the propagation of the shock. In this manner a stationary detonation wave is produced. Since the shock wave moves with supersonic velocity, the time from its first appearance to the onset of detonation is very short and an abrupt change occurs in the velocity of propagation. There exists a critical size of granular explosive charge above which the deflagration passes over to detonation and below which the explosive merely burns, first with increasing velocity, but, as the explosive is consumed, with decreasing velocity. The critical size depends on many factors such as the state of subdivision of the explosive, the pressure, the geometry and confinement of the explosive.

The chemical constitution of the explosive is important. For primary explosives the critical mass amounts only to a few milligrams while for high explosives it may amount to tons, and in the case of non-detonating explosives such as ammonium nitrate, to hundreds of tons. Some of the primary explosives form an exception to this since, as we have seen (see § 5.4) they detonate at the point of initiation.

### 5.6. Low-velocity detonation

The detonation stages observed in thin films of explosives are characterized by velocities varying from 1100 to 2500 m.sec.$^{-1}$ This value is much lower than the stable high-speed detonation normally observed in large charges (namely 4500–8000 m.sec.$^{-1}$). It may be identified with the *low* velocity detonation characteristic of nitroglycerine explosives and which has also been observed in

methyl nitrate, nitroglycerol and recently in solid explosives like
T.N.T. and tetryl. If the expansion of the gas product is taken into
account, this low-velocity detonation in thin films can be justified
on hydrodynamic grounds without postulating any detailed mechanisms for the propagation. (For a recent review of the hydrodynamic theory of detonation, see Eyring *et al.* (1949).)

By using the Abel equation of state,

$$P_2(V_2 - b) = n_2 R T_2, \tag{5.1}$$

and applying this equation to the reasoning in the Chapman-
Jouget hydrodynamic theory of detonation, the following expression
is obtained for the detonation velocity $D$ (Paterson, 1948):

$$D = \frac{V_1}{V_1 - b}(\gamma + 1) \bigg/ \sqrt{\left(\frac{n_2 R T_2}{\gamma}\right)}, \tag{5.2}$$

where $V_1$ is the volume of a gram of the explosive, $b$ is the covolume,
$n_2$ the number of moles of explosion products per gram, $R$ the molar
gas constant, $T_2$ the detonation temperature, $P_2$ the final pressure,
$V_2$ the final specific volume, and

$$\gamma = 1 + \frac{n_2 R}{C_2}. \tag{5.3}$$

Here $C_2$ is the specific heat of the products at constant volume.

This formula (5.2) applies only if no expansion of the products
takes place before the reaction is complete. *Bowden* and *Gurton*
(1949) have suggested that if an expansion does occur the net effect
on equation (5.2) is to increase the effective value of $V_1$ by a factor
depending on the amount of expansion, and to decrease $T_2$ by a
much smaller factor, so that if the expansion is large, as is probably
the case with thin layers of explosive, $V_1$ becomes so large that $b$ is
negligible in comparison. Then the term $\dfrac{V_1}{V_1 - b} \to 1$ and so equation
(5.2) becomes

$$D = \frac{\gamma + 1}{\gamma} \sqrt{(\gamma n_2 R T_2)}, \tag{5.4}$$

i.e.

$$D = \frac{\gamma + 1}{\gamma} C, \tag{5.5}$$

where $c$ is the velocity of sound in the products. Since $\dfrac{\gamma + 1}{\gamma} \simeq 2$
a reasonably stable minimum velocity should be obtained which is

about twice the velocity of sound in the products. Calculations for different explosives such as nitroglycerine and P.E.T.N. give velocities of $c.$ 2300 m.sec.$^{-1}$ in rough agreement with the observed value (see also Table XIX).

TABLE XIX. *Detonation velocities in thin films*

| Explosive | Velocity observed (m.sec.$^{-1}$) | Velocity calculated (m.sec.$^{-1}$) |
|---|---|---|
| Nitroglycerine | 2200 | 2240 |
| P.E.T.N. | 1450 | 2380 |
| Cyclonite | 2300 | 2340 |
| Tetryl | 1500 | 2050 |

*Propagation of the low-velocity detonation.* Some calculations have been made by Ratner (1947) on the temperature reached in a liquid explosive during high-velocity detonation ($c.$ 8000 m.sec.$^{-1}$). In the case of nitroglycerine the shock wave is sufficient to raise the temperature 3000° C. by adiabatic compression of the liquid itself. On the other hand, the low-velocity detonation (2000 m.sec.$^{-1}$) will only increase the temperature by some 40° C. and this is insufficient to maintain the explosive reaction. *Bowden* and *Gurton* have suggested that for liquid explosives, the mechanical action of the shock wave may facilitate the propagation. In other words, the shock wave front breaks up the liquid into small drops which are thrown into the reaction zone, and because of the large surface area, a rapid reaction results. As evidence for this view it has been shown that a change in viscosity has a marked effect on the ease of propagation. Thus it is not possible to detonate a thin film of the viscous diglycerol tetranitrate, and air-free nitroglycerine made viscous by the addition of a small amount of nitrocotton does not detonate at a low velocity.

When the explosive is present as a thin film between solid surfaces, its break-up into droplets may be facilitated by the separation of the surfaces. This is possible because the speed of sound in the solid is greater than the detonation velocity so that elastic waves in the metal due to the shock of the explosion will travel ahead of the detonation front. Some indication that standing waves might be set up in the metal which confines the explosive has been obtained in experiments where the blast marks of the explosion show an alternating pattern on the confining metal. This break-up of the explosive droplets also offers a reasonable explanation of the pitting and pock marks produced on brass during the detonation.

When the explosive is solid or is viscous this break-up is difficult and it is suggested that propagation is assisted by another mechanism. If air bubbles are present or are introduced into the liquid plastic or solid explosive, the adiabatic compression of these small gas bubbles by the shock front will produce local hot spots of several thousand degrees during the passage of the detonation wave. These hot spots will then become new explosion centres and will make propagation possible. In support of this view it is found that thin films of diglycerol tetranitrate will detonate only if air bubbles are introduced into the film of explosive. The phenomenon of 'dead pressing' in solid explosives, and the insensitiveness of old blasting gelatine and of cast explosives, would on this view be due to the removal of small air spaces. It is also supported by the observation that many explosives which propagate freely at atmospheric pressure fail to do so if the pressure is raised, for example, if the explosive is under water or at the bottom of a bore hole. If the initial pressure $P_1$ is high, the compression ratio $P_2/P_1$ is less and the adiabatic heating of the trapped gas is correspondingly reduced (see also Muraour (1950), and Basset and Muraour (1941)).

In high-melting primary explosives it is possible that the hot spots formed by intercrystalline friction in the detonation wave front may be sufficient to act as new explosion centres.

There is, therefore, evidence that the formation of hot spots (which has been shown to be the cause of initiation) is often the means by which propagation is rendered possible and by which the low-velocity detonation is maintained. Once again the commonest source of hot spots is the rapid compression of gas pockets, but in certain cases hot spots may be formed by intercrystalline friction, and they can also of course be formed by friction on grit particles. The general theory suggests that once the high-velocity detonation ($c.$ 8000 m.sec.$^{-1}$) sets in, it can continue to propagate without the aid of auxiliary hot spots.

PLATE III

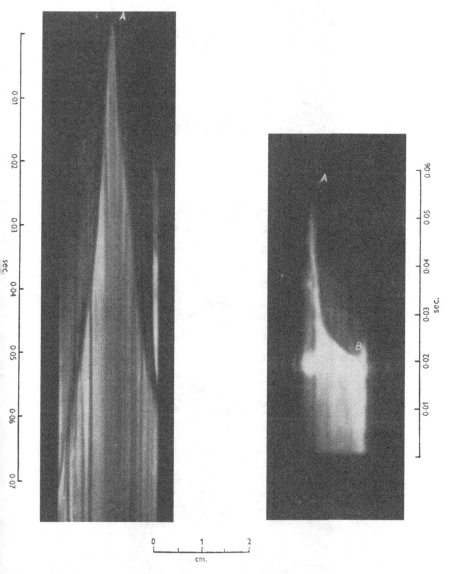

(a) Combustion of gunpowder in air. Initiation by hot wire at $A$. Streak photograph.

(b) Combustion of gunpowder under pressure showing acceleration of the burning and explosion at $B$. Streak photograph.

PLATE IV

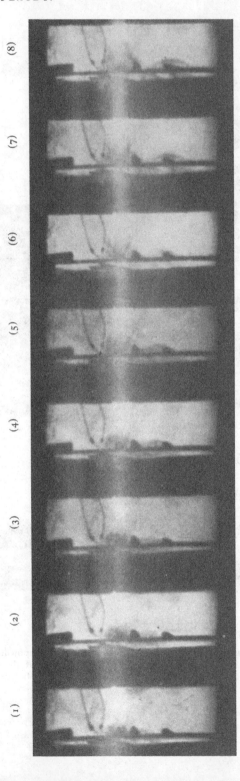

(1)  (2)  (3)  (4)  (5)  (6)  (7)  (8)

Burning of individual grains of gunpowder ignited by method shown in fig. 68. Sequence of pictures shows non-luminous reaction zone spreading from each grain until it ignites the next. The wire used for igniting the first grain is visible on the left. Each frame 1/1000 of a second (× 2). Silhouette photograph.

PLATE V

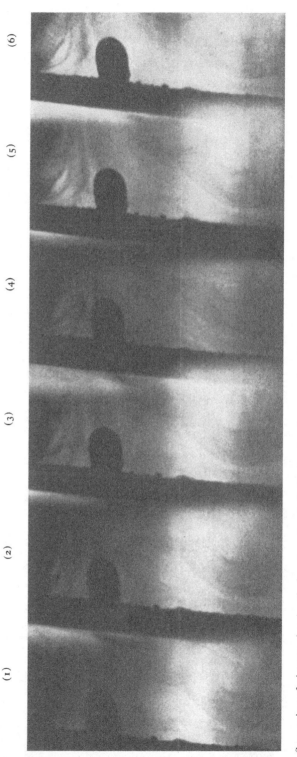

(1)   (2)   (3)   (4)   (5)   (6)

Succession of pictures showing the burning of an individual grain of gunpowder. The non-luminous zone about the grain gradually becomes luminous. Combustion of the grain occurs in a layer-wise fashion. Each frame 1/1000 of a second (× 10). Silhouette photograph.

## 5.7. The burning of some propellants

The combustion of propellants such as cordite and gunpowder occurs at a slow rate at atmospheric pressure, the speed of combustion being of the order of several millimetres per second. The speed increases with pressure according to fairly simple relations. In the case of cordite, investigators both in Great Britain and America consider that ignition begins in a reactive gas layer lying close to the surface of the cordite (see references by Kistiakowsky, 1948, and 'Symposium on kinetics of propellants arranged by the American Chemical Society, Sept. 1947).

Rideal and Robertson (1948 b) have evidence that when nitrocellulose, which forms the basis of cordite, is heated at atmospheric pressure a reactive mixture of gases consisting of aldehydes, nitrogen dioxide, nitric oxide, hydrogen and carbon monoxide are formed above the surface of the nitrocellulose, followed by partial liquefaction of the decomposing nitrocellulose. At a sufficiently high temperature (c. 180° C.) ignition of these reactive gases occurs and this is followed by a surface combustion of the nitrocellulose.

During the burning of cordite in a low-pressure atmosphere of nitrogen (e.g. 30 atm.) endothermic decomposition of the nitroglycerine and the nitrocellulose occurs at the surface giving nitrogen peroxide and volatile organic products, such as aldehydes (see § 3. 2. 9.). These gases react in the gas phase in a non-luminous zone close to the surface, and part of the heat of reaction is transferred back to the surface of the propellant by diffusion convection and radiation. The products of the gas phase reaction are nitric oxide, simple organic molecules, nitrogen, water, carbon monoxide and carbon dioxide. The temperature of this non-luminous zone rises to c. 1500° K. These gases are capable of further reaction at high temperatures and burn with a luminous flame at temperatures up to 3000° K. The observed increase in burning rate with pressure is probably due to two factors. The reaction in the gas phase is bimolecular, its rate increasing with pressure and so liberating heat more rapidly. The higher pressure also forces the luminous zone closer to the surface of the propellant, thus increasing the rate of heat transfer to the surface.

Kistiakowsky has suggested that the deflagration of crystalline explosives such as P.E.T.N., cyclonite and tetryl develops in the same way. There is first a surface decomposition giving volatile intermediates near the surface and this is followed by an exothermic reaction of these intermediates in the vapour phase. The heat liberated serves to heat the solid surface and so maintain the primary reaction (see also the work of Belajev (1938, 1939, 1941) and Andreev (1944)). As the external gas pressure increases the rate of the exothermic gas reaction and the rate of heating of the surface of the crystals is also increased. This results in an acceleration of the burning of the explosive.

*Gunpowder.* *Blackwood* and *Bowden* (1952) have recently obtained some drum camera photographs of the propagation of combustion in gunpowder. By means of a pulsed trigatron tube, which enables individual pictures to be taken at the rate of 2000 frames per second with a flash duration of 1 $\mu$sec., they have also been able to study the propagation of the combustion from grain to grain, and to conclude that the combustion probably occurs in a non-luminous reaction zone around the decomposing grain.

A typical photograph obtained for the combustion in air is given in Plate III *a*. Initiation is by a hot wire at *A* and the combustion accelerates slowly. The average rate of propagation in air at atmospheric pressure is of the order of 0·5 m.sec.$^{-1}$. A large amount of after-burning takes place in air (see Plate III*a*).

When the pressure is above atmospheric the propagation accelerates rapidly and passes over into explosion (Plate III*b*). The after-burning that occurs has the appearance of a gas explosion.

In Plate IV the decomposition of individual grains of the gunpowder is shown. Grains of gunpowder are placed several millimetres apart and one of the grains is ignited by a hot wire. A sketch of the apparatus used is given in fig. 68, showing the method for obtaining a simultaneous shadow record of the gunpowder grains and of the luminous flame from the burning cordite. The sequence of photographs in Plate IV shows that the grains burn in a layer-wise fashion and a non-luminous reaction zone spreads from each grain until it ignites the next grain. As the burning proceeds the reaction zone becomes luminous (Plate V). When this experiment is carried out under pressure, the grains disrupt and burn as a dust explosion with extreme rapidity.

## 5.8. Gaseous products of explosion

True detonation in high explosives normally gives as the major products gases such as $CO_2$, $CO$, $H_2O$, $N_2$ and $H_2$, the relative quantities obtained being in accordance with calculations based on equilibrium considerations. The thermal decomposition of explosives on the other hand gives more complicated products such as

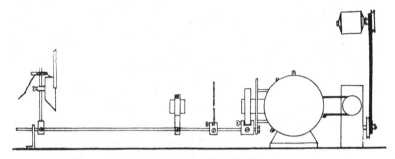

Sketch plan of camera and optical set up

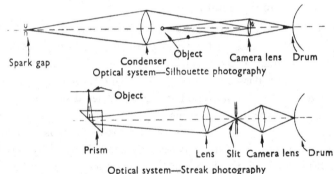

Optical system—Silhouette photography

Optical system—Streak photography

Fig. 68. Sketch of apparatus used for studying the burning of individual grains of gunpowder (see Plate III). The grains are placed in the light path ahead of the condenser lens.

nitrogen oxides, and aldehydes which may undergo further secondary reactions. The gases obtained from explosions initiated by impact have been analysed and compared with the products from detonation and thermal decomposition (Robertson and *Yoffe*, 1948) (see Table XX which gives some results for secondary explosives).

The actual composition of the gases liberated from impact explosions depends on the temperature and pressure reached under

TABLE XX. *Percentage composition of gases obtained from explosives*

|  | $NO_2$ | NO | $N_2O$ | $N_2$ | $CO_2$ | CO | $H_2$ | $O_2$ |
|---|---|---|---|---|---|---|---|---|
| (a) Nitroglycerine: | | | | | | | | |
| Detonation | — | — | — | 32·0 | 63·1 | — | — | 4·9 |
| Impact | — | 29·7 | 2·3 | 11·7 | 28·8 | 22·5 | 5·0 | — |
| Decomposition (180° C.) | — | 50·3 | 1·0 | 2·1 | 17·2 | 28·9 | 0·5 | — |
| (b) P.E.T.N.: | | | | | | | | |
| Detonation | — | 5·3 | — | 22·8 | 37·0 | 26·7 | 6·8 | 1·4 |
| Impact | — | 24·3 | 5·3 | 9·4 | 19·1 | 35·4 | 6·5 | — |
| Decomposition (210° C.) | 12·0 | 47·6 | 9·5 | 1·6 | 6·3 | 21·0 | 2·0 | — |
| (c) Cyclonite: | | | | | | | | |
| Impact | — | — | 11·3 | 43·3 | 18·8 | 21·3 | 5·3 | — |
| Decomposition (267° C.) | — | 22·5 | 22·8 | 31·0 | 13·2 | 8·7 | 1·8 | — |

the hammer, but it is clear that the composition of the gases approximates more closely to that of thermal decomposition than to the products formed by high-speed stable detonation. This observation is consistent with the view that the explosive is ignited at a hot spot and the propagation during the first stage is by a burning.

### 5.9. High-velocity detonation

It is clear from the experiments described in this chapter that the growth of the explosion from the point of initiation proceeds in stages. With many explosives it may begin as a slow burning which accelerates and passes over into a low-velocity detonation. It is suggested that the propagation of this is rendered possible by the presence of potential hot spots, e.g. by the presence of small pockets of air or gas between the explosive crystals. Further evidence in support of this has recently been obtained by *Bowden* and *Williams* (1951 b). Experiment shows that, if the initial gas pressure on a secondary explosive is increased to 50 atmospheres or so, the low-velocity detonation fails to propagate. Finally the high-velocity detonation, which is self-sustaining, sets in. The physics of this high-velocity detonation (Chapman, 1899; Jouguet, 1901) which is normally observed in large charges, is now reasonably well understood and its treatment is outside the scope of this monograph. The recent advances in the theory have been brought out in a Royal Society discussion which was led by W. G. Penny (1950), and contains contributions from G. I. Taylor (1950), H. Jones (1950) and others.

# REFERENCES

ALEXANDER, E. A. and LAMBERT, J. D. (1942). *Proc. Roy. Soc.* A, **179**, 499.

AMERICAN CHEMICAL SOCIETY (1947). Symposium on Kinetics of Propellants. Published 1950. *J. Phys. Colloid Chem.* **54**, 847 etc.

ANDREEV, K. K. (1934). *Chem. Zbl.* **1**, 489.

ANDREEV, K. K. (1935). *Bull. Soc. Chim. biol., Paris*, **2**, 2128.

ANDREEV, K. K. (1944). *C.R. Acad. Sci. U.R.S.S.* **44**, No. 1.

ANDREEV, K. K. (1946). *C.R. Acad. Sci. U.R.S.S.* **51**, 29, 123.

ANDREEV, K. K. and CHARITON, J. B. (1934). *Chim. et Industr.* **31**, 1040.

BASSET, J. and MURAOUR, H. (1941). *Chim. et Industr.* **45**, 218.

BELAJEV, A. F. (1938). *Acta Phys.-chim. U.R.S.S.* **8**, 763.

BELAJEV, A. F. (1939). *C.R. Acad. Sci. U.R.S.S.* **24**, No. 3.

BELAJEV, A. F. (1941). *Acta Phys.-chim. U.R.S.S.* **14**, 523.

BLACKWOOD, J. and BOWDEN, F. P. (1952). *Proc. Roy. Soc.* A, **213**, 285.

BOBOLEV, U. and CHARITON, V. (1937). *Acta Phys.-chim. U.R.S.S.* **7**, 416.

BOWDEN, F. P., EIRICH, F., MULCAHY, M. F. R., VINES, R. G. and YOFFE, A. D. (1943). *Bull. Coun. Sci. Industr. Res., Aust.*, no. 173.

BOWDEN, F. P. and GURTON, O. A. (1948a). *Nature, Lond.*, **161**, 348.

BOWDEN, F. P. and GURTON, O. A. (1948b). *Nature, Lond.*, **162**, 654.

BOWDEN, F. P. and GURTON, O. A. (1949). *Proc. Roy. Soc.* A, **198**, 337, 350.

BOWDEN, F. P., MULCAHY, M. F. R., VINES, R. G. and YOFFE, A. D. (1947). *Proc. Roy. Soc.* A, **188**, 291, 311.

BOWDEN, F. P. and RIDLER, K. E. W. (1936). *Proc. Roy. Soc.* A, **154**, 640.

BOWDEN, F. P., STONE, M. A. and TUDOR, G. K. (1947). *Proc. Roy. Soc.* A, **188**, 329.

BOWDEN, F. P. and WILLIAMS, H. T. (1951a). *Proc. Roy. Soc.* A, **208**, 176.

BOWDEN, F. P. and WILLIAMS, H. T. (1951b). *Research*, **4**, 339.

BOWDEN, F. P. and YOFFE, A. D. (1948). *Research*, **1**, 581.

CAIRNS, R. W. (1944). *Industr. Engng Chem.* **36**, 79.

CHAPMAN, D. L. (1899). *Phil. Mag.* **47**, 90.

CHARITON, J. B. and RATNER, S. B. (1943). *C.R. Acad. Sci. U.R.S.S.* **41**, 293.

CHERRY, T. (1945). *Rep. Coun. Sci. Industr. Res., Aust.*, A, **116**, No. 8.

COPP, J. L. and UBBELOHDE, A. R. (1948). *Philos. Trans.* A, **241**, 197.

COURTNEY-PRATT, J. S. (1949). *Research*, **2**, 287.

EGGERT, J. (1921). *Z. Elektrochem.* **27**, 547.

EIRICH, F. and TABOR, D. (1948). *Proc. Camb. Phil. Soc.* **44**, 566. See also TABOR, D. (1949). *Engineering*, **167**, 145.

EYRING, H., POWELL, R. E., DUFFEY, G. H. and PARLIN, R. B. (1949). *Chem. Rev.* **45**, 69.

FRANK-KAMENETSKI, D. A. (1939). *Acta Phys.-chim. U.R.S.S.* **10**, 365.

GARNER, W. E. (1938). *Trans. Faraday Soc.* **34**, 985, 1008.

GRAY, P. (1948). C.N.R.S. Colloquium, Paris. Conference on mechanism of inflammation and combustion processes in the gaseous phase. Paris.

GRAY, P. and YOFFE, A. D. (1949 a). *Research*, **2**, 339.

GRAY, P. and YOFFE, A. D. (1949 b). *Proc. Roy. Soc.* A, **200**, 114.

GRAY, P. and YOFFE, A. D. (1950). *J. Chem. Soc.* p. 3180.

HARGRAVE, K. K. (1947). *Trans. Faraday Soc.* Labile Molecule, p. 404.

HAWKES, A. S. and WINKLER, C. A. (1947). *Canad. J. Res.* **25**, 548.

HERZBERG, G. and WALKER, G. R. (1948). *Nature, Lond.*, **161**, 648.

JONES, E. (1928). *Proc. Roy. Soc.* A, **120**, 603.

JONES, E. (1949). *Proc. Roy. Soc.* A, **198**, 523.

JONES, H. (1950). *Proc. Roy. Soc.* A, **204**, 9.

JOST, W. (1946). *Explosion and Combustion Processes in Gases*. McGraw Hill.

JOUGUET, E. (1901). *C. R. Acad. Sci., Paris*, **132**, 673.

KALLMANN, H. and SCHRANKLER, W. (1933). *Naturwissenschaften*, **21-23**, 379.

KISTIAKOWSKY, G. B. (1948). *Third Symposium on Combustion and Flame and Explosion Phenomena*, p. 560. Madison, U.S.A.: Williams and Wilkins.

LAFITTE, P. (1925). *Ann. Phys., Paris*, **4**, 587.

LEWIS, B. and VON ELBE, G. (1939). *J. Chem. Phys.* **7**, 197.

MACDONALD, J. Y. (1938). *Trans. Faraday Soc.* **34**, 977.

MARINESCO, N. (1935). *C.R. Acad. Sci., Paris*, **201**, 1187.

MELDRUM, F. R. (1940). *Proc. Roy. Soc.* A, **174**, 410.

MULCAHY, M. F. R. (1948). *Phil. Mag.* **39**, 547.

MULCAHY, M. F. R. and VINES, R. G. (1947). *Proc. Roy. Soc.* A, **191**, 210, 226.

MURAOUR, H. (1933). *Chim. et Industr.* **30**, 39.

MURAOUR, H. (1938). *Trans. Faraday Soc.* **34**, 989.

MURAOUR, H. (1950). *Rev. Inst. Fr. Petrole*, **5**, 148.

MURAOUR, H. and WOHLGEMUTH, J. (1936). *Chim. et Industr.* **36**, No. 3.

PATERSON, S. (1948). *Research*, **1**, 221.

PATRY, M. (1933). Thesis (Nancy) Paris. (*See also* PATRY, M. and LAFITTE, P. (1931). *C.R. Acad. Sci., Paris*, **193**, 171, 1339.)

PAYMAN, W., SHEPHERD, W. C. F. and WOODHEAD, D. W. (1937). Safety in Mines Research Board Paper, No. 99.

PENNEY, W. G. (1950). *Proc. Roy. Soc.* A, **204**, 1.

PHILLIPS, L. (1947). *Nature, Lond.*, **160**, 753.

PHILLIPS, L. (1950). *Nature, Lond.*, **165**, 564.

RATNER, S. B. (1944). *C.R. Acad. Sci. U.R.S.S.* **42**, 265.

RATNER, S. B. (1947). *Acta Phys.-chim. U.R.S.S.* **22**, 357.

RICE, O. K. (1940). *J. Chem. Phys.* **8**, 727.

RIDEAL, E. K. and ROBERTSON, A. J. B. (1948 a). *Proc. Roy. Soc.* A, **195**, 135.

RIDEAL, E. K. and ROBERTSON, A. J. B. (1948 b). *Third Symposium on Combustion and Flame and Explosion Phenomena*, p. 536. Madison, U.S.A.: Williams and Wilkins.

ROBERTSON, A. J. B. (1947). Thesis for Ph.D., Cambridge.

ROBERTSON, A. J. B. (1948 a). *Trans. Faraday Soc.* **44**, 977.

ROBERTSON, A. J. B. (1948 *b*). *J. Soc. Chem. Ind.* **67**, 221.
ROBERTSON, A. J. B. and YOFFE, A. D. (1948). *Nature, Lond.*, **161**, 806.
ROGINSKY, S. (*See* SEMENOFF, N.)
SCHMID, O. (1940). *Z. Phys. Chem.* A, **186**, 113.
SEMENOFF, N. (1935). *Chemical Kinetics and Chain Reactions.* Oxford.
STOUT, H. P. (1950). *Nature, Lond.*, **166**, 28.
SULTANOFF, M. (1950). *Rev. Sci. Instrum.* **21**, 653.
TAYLOR, G. I. (1950). *Proc. Roy. Soc.* A, **204**, 8.
TAYLOR, W. and WEALE, A. (1932). *Proc. Roy. Soc.* A, **138**, 92.
TAYLOR, W. and WEALE, A. (1938). *Trans. Faraday Soc.* **34**, 995.
THOMAS, P. H. (1949). Thesis for Ph.D., Cambridge.
UBBELOHDE, A. R. (1948). *Philos. Trans.* A, **241**, 197.
UBBELOHDE, A. R. (1950). *Research*, **3**, 207.
UBBELOHDE, A. R. and WOODWARD, P. (1948). *Philos. Trans.* A, **241**, 222.
VAN'T HOFF, J. H. (1884). *Études de Dynamique Chimique.*
VAUGHAN, J. and PHILLIPS, L. (1949). *J. Chem. Soc.* p. 2741.
VINES, R. G. (1947). *Nature, Lond.*, **160**, 400.
YOFFE, A. D. (1948). *Nature, Lond.*, **161**, 349.
YOFFE, A. D. (1949). *Proc. Roy. Soc.* A, **198**, 373.
YOFFE, A. D. (1951). *Proc. Roy. Soc.* A, **208**, 188.

## General References

Important contributions to the mechanism of the decomposition of the azides have been made by W. E. Garner and co-workers at Bristol. See, for example, Garner, W. E. and Maggs, J. (1939), *Proc. Roy. Soc.* A, **172**, 299. A theoretical treatment of this has been given by Mott, N. F. (1939), *Proc. Roy. Soc.* A, **172**, 325. A survey of work before 1934 has been given by Audrieth, L. F. (1934), *Chem. Rev.* **15**, 169.

# INDEX

Adiabatic compression, 3
  effect of external conditions on, 56, 57, 58, 61
  importance in initiation, 33
  and nature of gas, 35–8
air pockets, *see* gas bubbles
alkyl nitrates, glow of, 50
area of contact, 12
azides,
  effect of confinement, 87
  influence of their structure on detonation velocity, 87; on sensitivity, 87

Blast patterns, information from, 74
blasting gelatine, sensitized by gas bubbles, 32
butane-diol-dinitrate, sensitized by gas bubbles, 32
butane-triol-trinitrate, sensitized by gas bubbles, 32

Cadmium fulminate, sensitivity of, 86
calcium azide, growth of explosion in, 87
  sensitivity of, 86
cameras, image converter, 68–70
  rotating drum, 67–8
cavity, explosion in, 70–7
  striker, 30, 37, 42
cordite, burning of, 93
critical size, 88–9
cuprous-fulminate, sensitivity of, 86
cyanuric triazide, 87
cyclonite,
  products in thermal decomposition, 96
  sensitized by grit, 22, 23, 24
cyclonite explosion
  growth of, 83
  influenced by grit, 21
  initiation by friction, 21; by impact, 61

Dead pressing, 92
decomposition products, comparison of detonation, impact and thermal, 96

delay-time of explosion, with liquid explosives, 42–8; *see also* induction period
  with solid explosives, 61–3
detonation,
  critical size for, 88
  in cyclonite, 83
  development of, 76
  in diglycerol tetranitrate, 92
  high-velocity, 96
  hydrodynamic theory of, 90
  in lead azide, 86
  in lead styphnate, 86
  low velocity, 89, 91
  mechanism suggested, 88
  in mercury fulminate, 86
  in nitroglycerine, 77–82
  in P.E.T.N., 83
  products of, 95–6
  in tetryl, 83
  in thin films, 89–91
diethylene glycol dinitrate, growth of explosion in, 82
diglycol dinitrate, sensitized by gas bubbles, 32
diglycol tetranitrate, growth of explosion in, 82
  sensitized by gas bubbles, 32

Ethyl nitrate, propagation of, 49
  sensitization of, by gas bubbles, 32
explosion,
  catalyst, 5, 6
  efficiency, 29
  growth of, 10, 70
  initiation of, 4; by friction, 21–2; by nuclear particles 10
  measurement of transient pressure in, 38; nucleus, 5; size of, 9; patterns, 74–7; prevention of, 5; propagation velocity of, 4
  products of, 93, 95–6
  radicals in, 6, 50
  thermal, 5–9
  transient temperature for, 26
explosives, *see also* individual explosives
  decomposition of, at high dilution, 49

102 INDEX

explosives (cont.)
distribution of, 31, 45, 56
flow of, under impact, 59–60
glow of, 50
impact sensitivity of liquid and solid, 32
nature of initial burning in, 88
primary, 23, 25; growth of explosion in, 86; initiation by impact, 65
secondary, 21; sensitivity to shearing, 21–3
velocity constants of decomposition of, 27

Flow of solid explosives, 59–60
friction, initiation of explosives, 17

Gamma, influence of, 35, 37, 38, 45
variation with compression time, 47
gas bubbles,
adiabatic compression of, 33
calculation of temperature rise, 33, 39–42
danger of, 32
heat developed in, 42
influence of gamma, 35, 37, 38; of pressure, 34–9
initiation by, 29, 33
pressure rise of, 33, 38–9
size of, 29
temperature rise of, 33, 39–42
trapping of, 29–33, 60
gas pockets, importance in propagation, 89–92, 96
grit acting on the explosion, 2, 19
hardness of, 19, 25, 26, 63
melting point of, 19–20, 63, 66
size of, 25, 63, 64, 65
thermal conductivity of, 25
growth of explosion,
in liquids, 70–82
in solid primary explosives, 86–7
in solid secondary explosives, 82–5
gunpowder,
burning of individual grains, 94–5
explosion in, initiation by gas pockets 66; by grit, 66

Hexanitrosobenzene explosion of, 87
hot spots,
critical temperatures of, 65
demonstration of, 20
duration of, 13, 14, 26
influence of melting point on, 18

influence of sliding speed on, 14
importance of, in initiation, 1
and load, 15
limiting temperature for, 12, 14, 18, 19
measurement by infra-red cell, 14
methods of formation of, 1
photographing of, 14
produced by intercrystalline friction, 92; by shock waves, 92
in the presence of liquid, 15
simultaneous measurement of explosion, 19, 20
size of, 14, 64–5
temperature calculation of, 13
temperature necessary for explosion, 18, 39–42; for initiation, 26
and thermal conductivity, 14, 15–16
visual observation of, 13, 14, 16
hydrogen peroxide-methyl alcohol mixtures, sensitization by gas bubbles, 32

Ignition temperatures, 60
impact, time of, 39–41, 74
impact apparatus, 28
impact initiation by gas bubbles, 29
impact initiation of solids,
distribution of explosive for, 56
hardness of striker for, 58–9
influenced by pressure, 57–8
impact sensitivity, influenced by pressure, 33–42, 61
induction period, 26, 60
of lead azide, 9
initial burning,
nature of, 88
of liquid explosives, 70–82
of solid primary explosives, 86–7
of solid secondary explosives, 82–5
initiation
by electrostatic charge, 11
by friction, 12, 17, 21, 22
by friction, influence of thermal conductivity, 17–18
by hot wire, 86, 94
by impact of primary explosives, 65
by inter-crystalline friction, 56, 92
in liquids, 17–19
in liquids by entrapped gas, 29–42
in solids, 19–27
of solids by entrapped gas, 56–7
by spark 77, 82
by supersonics, 11
time for, 61, 62, 66